Daniel Jordan

Agricultural statistics of Ireland

1889

Daniel Jordan

Agricultural statistics of Ireland
1889

ISBN/EAN: 9783742800244

Manufactured in Europe, USA, Canada, Australia, Japa

Cover: Foto ©Klaus-Uwe Gerhardt /pixelio.de

Manufactured and distributed by brebook publishing software
(www.brebook.com)

Daniel Jordan

Agricultural statistics of Ireland

THE

AGRICULTURAL STATISTICS

OF

IRELAND,

FOR THE YEAR

1889.

DIVISION OF LAND: ACREAGE UNDER CROPS;
NUMBER AND SIZE OF HOLDINGS; RATES OF PRODUCE;
NUMBER AND AGES OF LIVE STOCK;
EXPORTS AND IMPORTS OF LIVE STOCK; HONEY PRODUCED;
NUMBER OF SCUTCHING MILLS; SILOS AND ENSILAGE;
THE WEATHER.

Presented to both Houses of Parliament by Command of Her Majesty.

DUBLIN:
PRINTED FOR HER MAJESTY'S STATIONERY OFFICE
BY
ALEXANDER THOM & CO. (LIMITED),
And to be purchased, either directly or through any Bookseller, from
EYRE and SPOTTISWOODE, East Harding-street, Fetter-lane, E.C., or 32, Abingdon-street, Westminster, S.W.; or ADAM and CHARLES BLACK, 6, North Bridge, Edinburgh;
or HODGES, FIGGIS, and Co., 104, Grafton-street, Dublin.

CONTENTS.

Page

INTRODUCTORY REMARKS:—

PART I.—TILLAGE; MEADOW AND CLOVER, &c.

TABLE I.—Acreage under Crops in 1858 and 1859, with proportionate Area under each Crop, . . . 4

II.—Extent of Land and proportionate Area under Crops, Grass, Fallow, Woods and Plantations, and Bog, Waste, Water, &c., in each year from 1849 to 1859, . . . 6

" III.—Number of Holdings, by Classes, in each County and Province, in 1858 and 1859, . . . 8

" IV.—Approximate Return of the Number of Occupiers resident in each County and Province in 1859, classified according to the total extent of land held, . . . 9

V.—Number of Holdings in each Province in 1841, 1851, 1853, 1857, 1858, and 1859, according to the classification of the Census Commissioners of 1841, . . . 10

PART II.—PRODUCE OF THE CROPS.

CHARACTER [illegible] THE PRODUCE OF THE CROPS [illegible] WHEAT, OATS, [illegible], . . . 11

TABLE VI.—Total Produce of the principal Crops in 1858 and 1859, and the Increase or Decrease in the latter year, . . . 12

VII.—Estimated Average Produce per Statute Acre of the principal Crops in 1858 and 1859, and the Increase or Decrease in the latter year, . . . 13

VIII.—Extent under each of the principal Crops in Statute Acres, the Total Produce, and the Average Yield per Acre, for each year from 1850 to 1859, . . . 14

PART III.—LIVE STOCK.

IX.—Number and Ages of Live Stock in 1858 and 1859, and the Increase or Decrease in each description for the latter year, . . . 15

" X.—Number of Live Stock in each year from 1850 to 1859, . . . 15

" XI.—The proportion per cent. of Horses, Cattle, Sheep, and Pigs according to Ages, from 1850 to 1859, . . . 15

" XII.—Number of Milch Cows in each year from 1854 to 1859, . . . 16

Exports and Imports of Live Stock, . . . 16

Honey produced in 1859, . . . 17

Number of Scutching Mills, . . . 17

Flax and Tillage, . . . 19

SUMMARY TABLES.

TILLAGE; MEADOW AND CLOVER; &c.

TABLE 1.—Number of Holdings, their Size in Statute Acres, and the Division of Land in each County and Province in 1859, . . . 20

" 2.—Proportion per cent. of Total Area under Crops, Grass, Fallow, Woods and Plantations, Bog and Marsh, Barren Mountain Land, and Water, Roads, Fences, &c., in each County and Province, . . . 20

" 3.—Number of Holdings and Size in Statute Acres, and the Division of Land in 1859 by Poor Law Unions, . . . 21

" 4.—Proportion per cent. under Crops, Grass, Fallow, &c., by Poor Law Unions, . . . 22

" 5.—Extent of Land under Crops in 1859, Valuation in 1859, and Population in 1851, by Counties and Provinces, . . . 24

" 6.—Produce of the Crops in 1859, by Counties and Provinces, . . . 25

" 7.—Extent of Land under Crops in 1859, Valuation in 1859, and Population in 1851, by Poor Law Unions, . . . 26

" 8.—Produce of the Crops in 1859, by Poor Law Unions, . . . 29

" 9.—Number of Holdings exceeding One Acre, the extent of Land under Crops in each year, from 1850 to 1859, by Counties and Provinces, . . . 32

" 10.—Average Rates of Produce of Crops per Statute Acre in each year from 1850 to 1859, by Counties and Provinces, . . . 41

LIVE STOCK.

" 11.—Number of Stockholders, and Quantity of Live Stock in 1859, by Counties and Provinces, . . . 46

" 12.—Number of Stockholders and Quantity of Live Stock in 1859, by Poor Law Unions, . . . 47

" 13.—The Quantity of Live Stock in each year from 1850 to 1859, by Counties and Provinces, . . . 51

" 14.—Total Area under Potatoes, and the Extent in Statute Acres under each description of that Crop planted in 1859, by Counties and Provinces, . . . 55

" 15.—Total Area under Potatoes and the Extent planted of each description of that Crop in 1859, by Poor Law Unions, . . . 57

" 16.—The Average Rate of Produce per Acre of each description of Potato planted in Ireland in 1859, by Counties, . . . 60

OBSERVATIONS of District-Inspectors of the Royal Irish Constabulary, and of Sergeants of the Metropolitan Police, as the probable amount of the good or bad yield of the Crops in each of their Districts, . . . 60

APPENDIX.

DETAILS REGARDING SALES and EXCHANGES furnished by Owners and Occupiers of Land, . . . 79

ABSTRACT of the METEOROLOGICAL OBSERVATIONS registered at the Ordnance Survey Office, Phoenix Park, Dublin, . . . 104

Remarks on the Weather of the year 1859, by J. W. Moore, Esq., [illegible], . . . 124

OBSERVATIONS

OF THE

AGRICULTURAL STATISTICS OF IRELAND,

FOR THE YEAR 1880.

TO HIS EXCELLENCY LAWRENCE, EARL OF ZETLAND,

&c., &c., &c.,

LORD LIEUTENANT-GENERAL AND GENERAL GOVERNOR OF IRELAND.

MAY IT PLEASE YOUR EXCELLENCY,

I have the honour to present to your Excellency the following Report and detailed Tables concerning Agriculture in Ireland for the year 1889, which have been compiled and arranged in the same manner as those for the previous year.

A review of the detailed Tables confirms the observations I made when presenting the General Abstracts in August last, and the Produce Returns in January of this year.

The following is an analysis of the information contained in the tables:—

PART I.—TILLAGE; MEADOW AND CLOVER; &c.

The acreage under Crops, Grass, Fallow, Woods and Plantations, and Bog, Waste, Water, &c., in 1888 and 1889, was as follows:—

—	1888.	1889.	Increase or Decrease between 1888 and 1889.	
			Increase.	Decrease.
	Acres.	Acres.	Acres.	Acres.
Under Crops, including Meadow and Clover,	5,140,653	5,056,014	—	84,667
„ Grass, or Pasture,	9,906,097	9,999,297	93,200	—
„ Fallow,	15,613	12,450	—	3,163
„ Woods and Plantations,	331,587	326,636	—	4,951
„ Bog, Waste, Water, &c.,*	4,935,773	4,935,354	—	419
Total,†	20,328,753			

The area under Crops in 1889, compared with 1888 shows a net decrease of 84,667 acres—there being a decrease of 50,209 acres in tillage, and of 76,968 acres in the area under hay on permanent pasture or grass not broken up in rotation; while there is an increase of 42,310 acres under hay on clover, sainfoin, and grasses under rotation. There is an increase of 93,200 acres under Grass, while there is a decrease of 3,163 acres of Fallow land, of 4,951 acres under Woods and Plantations, and of 419 acres under Bog, Waste, Water, &c.

Of the 4,935,354 acres given as under "Bog, Waste, Water, &c.," in 1889, 1,770,493 acres were enumerated as "Bog and Marsh," 2,275,733 acres as "Barren Mountain Land," and 889,128 acres as "Water, Roads, Fences, &c." Compared with 1888 "Bog and Marsh" appears to have decreased by 1,957 acres, "Barren Mountain Land" by 3,238 acres, and "Water, Roads, Fences, &c.," shows an increase of 4,776 acres.

The area and proportionate extent of each crop in 1888 and 1889, with the increase or decrease in the latter year, are given in the following Table (I.), from which it appears that, compared with 1888, there was last year a net decrease of 35,852 acres in cereals, wheat having decreased by 9,365 acres, oats by 41,906 acres, and beans and pease by 1,467 acres, while barley increased by 14,854 acres, and bere and rye by 1,985 acres.

In green crops there was a net decrease of 14,396 acres, potatoes having decreased by 17,332 acres, mangel wurzel by 1,729 acres, vetches and rape by 1,973 acres, and cabbage by 71 acres, while turnips increased by 3,676 acres, and carrots, parsnips, and other green crops by 3,089 acres.

Flax shows an increase of 39 acres, and meadow and clover a decrease of 84,468 acres.

In 1889, 30·4 acres in every 100 under crops were under cereals, 24·1 under green crops, 2·2 under flax, and 43·3 under meadow and clover.

* Including [illegible] acres under water. † Exclusive of [illegible] acres under the larger rivers, lakes, and tideways.

Varieties of Potatoes.

POTATOES.—The tables relating to the potato crop point to several important conclusions. It will be observed (See Table 14, p. 56) that of the 787,234 acres planted with potatoes, 79·6 per cent. belonged to one variety, namely, "Champions," showing no appreciable difference in the percentage of this variety as compared with the previous year. Of the total area under potatoes 7·4 per cent. was under Flounders, 3·7 per cent. under Skerry Blues, 2·0 per cent. under White Rocks, 1·9 per cent. under Magnum Bonums, 1·1 per cent. under Scotch Downs, 1·0 per cent. under Kemps, and 3·4 per cent. under all other varieties. It will be seen by a reference to Table 16 that not only was the Champion variety the one planted in greatest quantity, but that it was generally the most prolific in its yield.

Table 16 also points out the best potato-growing districts in Ireland, and the varieties which appear to thrive best in particular counties.

Extent under Crops.

Of the total extent under crops in 1889, 83·4 per cent., or over four-fifths, were under three crops—oats (24·5), potatoes (15·6), and meadow and clover (43·3).

(TABLE I.)—The Acreage under Crops in 1888 and 1889, and the Increase or Decrease in the latter year:—

The extent of land under grass in 1889 (*exclusive of that under meadow and clover*) was 9,996,297 acres, or 49·2 in every 100 of the entire country, against 9,903,057 acres or 48·7 per cent. in 1888. The relative proportions under grass in each Province were—in Munster 53·3 per cent. in 1889, and 53·1 per cent. in 1888; Leinster 58·9 per cent. in 1889, and 58·0 per cent. in 1888; Connaught 48·4 per cent. in 1889, and 47·1 per cent. in 1888; and Ulster 40·9 per cent. in 1889, and 41·3 per cent. in 1888. **Grazing Land, 1888 and 1889.**

There appears to have been an increase of pasture land in 1889 in Leinster of 0·9 per cent. of the total area of the province, in Munster of 0·2 per cent., and in Connaught of 1·3 per cent.; while there has been a decrease of 0·4 per cent. in Ulster.

Of the counties—Limerick, Meath, and Westmeath had each above 60 acres in every 100 of their entire area under grass in 1889; Clare, Fermanagh, Kildare, Kilkenny, Roscommon, and Tipperary had above 55 and under 60 acres; Carlow, Cavan, Cork, Leitrim, Longford, Queen's, Sligo, Waterford, and Wexford had from 50 to 55 acres; Antrim, Dublin, Galway, Kerry, King's, Louth, Mayo, Monaghan, Tyrone, and Wicklow had above 40 and under 50 acres; and Armagh, Donegal, Down, and Londonderry, had over 30 and under 40 acres in every 100 acres under grass in 1889. Only 32·6 per cent. of the total **area** of Donegal was enumerated in 1889 as under grass, while Meath shows the highest percentage, 70·1. **Grazing Land in 1889.**

The area of each County and Province, **and the extent and percentage under grass** in 1889, are given at page 30.

Of the total area of Ireland (20,824,708 statute acres),* the land under grass in 1889 was, as already stated, nearly one-half. It appears from the succeeding Table (II.) to have decreased from 50·4 per cent. of the total area in 1880 to 48·2 in 1885, but during the ten years the proportion of grass varied from 50·6 per cent. in 1884 to 48·7 in 1888. **Division of Land, 1880–1889.**

In Crops a decrease took place from 5,061,084 acres or 25·0 per cent. of the total area in 1880 to 4,872,744 acres, or 24·0 per cent. in 1884, from which date until 1888 there was annually a slight increase, but in 1889 there was a decrease, the extent being 5,056,016, or 24·8 per cent. of the total area, against 25·3 per cent. in 1888.

Fallow or uncropped arable land amounted to 15,400 acres in 1880, and to 12,450 acres in 1889.

Comparing the first and the last years of the **decade, Woods and Plantations exhibit** a decrease from 332,258 acres to 328,485 acres.

In "Bog, Waste, Water, &c." an increase is shown—from 4,633,297 acres in 1880, **to** 4,936,354 acres in 1889, the difference being equivalent to 1·5 per cent. of the total area.

TABLE II.—The Extent of Land in Statute Acres, and the proportionate Area, under Crops, Grass, Fallow, Woods and Plantations, and Bog, Waste, Water, &c., in each Year from 1880 to 1889, also the Number of Holdings exceeding 1 Acre:—

Years.	Number of Holdings exceeding 1 Acre.	Extent of Land in Statute Acres under: Crops (including Meadow and Clover).	Grass.	Fallow.	Woods and Plantations.	Bog, Waste, Water, &c.	Total.*	Percentage per Cent. under: Crops (including Meadow and Clover).	Grass.	Fallow.	Woods and Plantations.	Bog, Waste, Water, &c.
		Acres.	Acres.	Acres.	Acres.	Acres.						
1880	[illegible]	5,061,084	10,[illegible]	[illegible]	[illegible]	4,633,297	20,824,708 acres.	25·0	50·4	[illegible]	1·7	22·8
1881	[illegible]	[illegible]	[illegible]	[illegible]	[illegible]	[illegible]		[illegible]	50·0	[illegible]	[illegible]	[illegible]
1882	[illegible]	[illegible]	[illegible]	[illegible]	[illegible]	[illegible]		[illegible]	[illegible]	[illegible]	[illegible]	[illegible]
1883	[illegible]	[illegible]	[illegible]	[illegible]	[illegible]	[illegible]		[illegible]	[illegible]	[illegible]	[illegible]	[illegible]
1884	[illegible]	4,872,744	[illegible]	[illegible]	[illegible]	[illegible]		24·0	[illegible]	[illegible]	[illegible]	[illegible]
1885	[illegible]	[illegible]	[illegible]	[illegible]	[illegible]	[illegible]		[illegible]	[illegible]	[illegible]	[illegible]	[illegible]
1886	[illegible]	[illegible]	[illegible]	[illegible]	[illegible]	[illegible]		[illegible]	[illegible]	[illegible]	[illegible]	[illegible]
1887	[illegible]	[illegible]	[illegible]	[illegible]	[illegible]	[illegible]		[illegible]	[illegible]	[illegible]	[illegible]	[illegible]
1888	[illegible]	[illegible]	[illegible]	[illegible]	[illegible]	[illegible]		[illegible]	48·7	[illegible]	[illegible]	[illegible]
1889	[illegible]	5,056,016	9,996,297	[illegible]	[illegible]	4,936,354		24·8	[illegible]	[illegible]	[illegible]	[illegible]

Barren Mountain Land, 1859. "Barren Mountain Land" covers an area of 100,000 acres and upwards in the following seven counties, viz.:—Donegal, 827,342 acres, or 27·6 per cent. of its entire area; Kerry, 298,517 acres, or 25·7 per cent.; Galway, 235,251 acres, or 15·7 per cent.; Cork, 255,468 acres, or 13·9 per cent.; Mayo, 203,484 acres, or 15·4 per cent.; Tyrone, 113,080 acres, or 14·5 per cent.; and Wicklow, 120,229 acres, or 24·1 per cent.

14·5 per cent. of Sligo, or 65,428 acres, 7·2 per cent., or 77,087 acres of Tipperary, and 17·3 per cent., or 79,893 acres of Waterford are under "Barren Mountain Land." The counties containing the smallest areas under "Barren Mountain Land" are Meath with 632 acres, or 0·1 per cent. of its entire area; Longford, 1,674 acres, or 0·6 per cent.; Westmeath, 530 acres, or 0·1 per cent.; Kildare, 2,297 acres, or 0·6 per cent.; and Monaghan, 4,922 acres, or 1·5 per cent. Only 217,861 acres, or 4·5 per cent. of Leinster are returned as being under "Barren Mountain Land," while 819,429 acres, or 13·8 per cent. of Munster; 686,837 acres, or 12·9 per cent. of Ulster; and 549,606 acres or 13·0 per cent. of Connaught are so returned.*

Water, Roads, Fences, &c., 1859. Very little variation is exhibited in the proportionate area under "Water, Roads, Fences, &c." in the several counties and provinces. In the counties the highest percentage is 7·5 in Dublin, and the lowest 3·2 in Roscommon and Wicklow. 889,198 acres (including 153,035 acres under water), or 4·4 per cent. of the entire area of the country, were returned in 1859 as "Water, Roads, Fences, &c." This, however, does not include the acreage under the larger rivers, lakes and tideways. See note (†), page 2.

A table showing the division of land by Poor Law Unions is given at pages 21 and 23.

Number and size of holdings, 1858 and 1859. According to the returns for 1859, the number of separate holdings was 565,973, being 3,383 more than in the previous year. The holdings which increased in number were—those "not exceeding 1 acre" by 1,978; those "above 1 and not exceeding 5 acres" by 1,334; "above 5 and not exceeding 15 acres" by 415; those "above 50 and not exceeding 100 acres" by 44; those "above 100 and not exceeding 200 acres" by 129; and those "above 500 acres" by 24. The holdings which decreased in number were those "above 15 and not exceeding 30 acres" by 215; those "above 30 and not exceeding 50 acres" by 361; and those "above 200 and not exceeding 500 acres" by 5.

Size of Holdings.	Number in 1858.	Number in 1859.	Increase or Decrease in 1859.	
			Increase.	Decrease.
Not exceeding 1 Acre,	47,851	49,829	1,978	–
Above 1 and not exceeding 5 Acres,	60,256	61,590	1,334	–
„ 5 „ 15 „	156,146	156,561	415	–
„ 15 „ 30 „	135,311	135,096	–	215
„ 30 „ 50 „	73,763	73,403	–	361
„ 50 „ 100 „	58,476	58,520	44	–
„ 100 „ 200 „	22,796	22,925	129	–
„ 200 „ 500 „	8,372	8,367	–	5
Above 500 Acres,	1,561	1,585	24	–
Total,	563,045	565,973	3,333	–

* With reference to the question whether waste land is increasing or decreasing in Ireland, the following from a Paper read by Dr. Grimshaw before the Statistical and Social Inquiry Society of Ireland on the 17th of April, 1884, may be of interest:—

"The following Table shows that so far from the waste land of Ireland being on the increase, an immense amount of waste land has been reclaimed during the past forty years.

"Division of Land in 1841, '51, '61, '71, and '81.

Division of Land.	1841.	1851.	1861.	1871.	1881.
	Statute Acres.	Statute Acres.	Statute Acres.	Statute Acres.	Statute Acres.
Under Crops (including Meadow),	[illegible]	[illegible]	[illegible]	[illegible]	[illegible]
Grass,		[illegible]	[illegible]	[illegible]	[illegible]
Woods and Plantations,	[illegible]	[illegible]	[illegible]	[illegible]	[illegible]
Barren Mountain Land,	[illegible]	[illegible]	[illegible]	[illegible]	[illegible]
Bog and Marsh,					[illegible]
Waste Land, &c.,					[illegible]
Total,	[illegible]				

A table showing the number of holdings, by classes, for each Poor Law **Union, in 1889,** will be found on pp. 21 and 22.

The number of separate holdings in each county and province, in 1888 and 1889, is given by classes in Table III., at page 8.

As in many instances landholders occupy more than one farm, and as, in other cases, farms extend into two or more townlands—the portion in each townland being enumerated and classified as a separate holding—it has been considered desirable, with the view of ascertaining the number of Occupiers, and of classifying them according to the total extent of land held by each, to obtain a Return of the number of persons having more than one farm or holding. Each Enumerator is, therefore, required to furnish the name of every landholder residing in his district who has two or more farms, or whose farm extends into two or more townlands, together with the area of each portion, and the locality in which it is situated. The number of actual occupiers in 1889 thus arrived at is given in Table IV., page 9, by counties and provinces. On comparing the results in this Table with the figures given in Table III., it appears that in 1889 there were 585,875 holdings in the hands of 525,152 occupiers. Number of separate Holdings and of Occupiers, 1888 and 1889.

The number of separate holdings and the number of occupiers in **each Province in** 1888 and 1889 were:—

Provinces.	Number of Separate Holdings.		Number of Occupiers.	
	1888.	1889.	1888.	1889.
Leinster, . .	130,884	130,864	108,281	108,791
Munster, . .	132,267	135,465	111,104	112,304
Ulster, . .	188,779	200,884	181,460	186,418
Connaught, . .	131,540	131,208	114,429	114,609
Total, . .	582,549	585,875	521,465	525,152

The number **of occupiers of land in 1889 was 525,152, being 3,687 more than in** the previous year.

Excluding those holding land "not exceeding one acre," who are to a great extent merely occupiers of small gardens, they numbered 475,817 in 1889, or 1,702 more than in 1888. There was a decrease in Leinster of 34—from 92,751 in 1888 to 92,717 in 1889; but an increase in Munster of 319—from 99,442 in 1888 to 99,761 in 1889; in Ulster of 1,182—from 172,889 in 1888 to 174,071 in 1889; and in Connaught of 255—from 109,053 in 1888 to 109,308 in 1889. The increase in occupiers holding land above 1 and not exceeding 30 acres was 1,308 and the number holding land exceeding that acreage increased by 467.

TABLE III.—The number of Holdings, by classes, for each County and Province, in 1888 and 1889, and the increase or decrease in the latter year:—

COUNTIES.	Year	Not exceeding 1 Acre.	Above 1 and not exceeding 5 Acres.	Above 5 and not exceeding 15 Acres.	Above 15 and not exceeding 30 Acres.	Above 30 and not exceeding 50 Acres.	Above 50 and not exceeding 100 Acres.	Above 100 and not exceeding 200 Acres.	Above 200 and not exceeding 500 Acres.	Above 500 Acres.	Total.
		Number and Classification of Holdings.									
ANTRIM,	1888	[illegible]	[illegible]	[illegible]	[illegible]	[illegible]	[illegible]	[illegible]	[illegible]	[illegible]	[illegible]
	1889	[illegible]	[illegible]	[illegible]	[illegible]	[illegible]	[illegible]	[illegible]	[illegible]	[illegible]	[illegible]
ARMAGH,	1888	[illegible]	[illegible]	[illegible]	[illegible]	[illegible]	[illegible]	[illegible]	[illegible]	[illegible]	[illegible]
	1889	[illegible]	[illegible]	[illegible]	[illegible]	[illegible]	[illegible]	[illegible]	[illegible]	[illegible]	[illegible]
CARLOW,	1888	[illegible]	[illegible]	[illegible]	[illegible]	[illegible]	[illegible]	[illegible]	[illegible]	[illegible]	[illegible]
	1889	[illegible]	[illegible]	[illegible]	[illegible]	[illegible]	[illegible]	[illegible]	[illegible]	[illegible]	[illegible]
CAVAN,	1888	[illegible]	[illegible]	[illegible]	[illegible]	[illegible]	[illegible]	[illegible]	[illegible]	[illegible]	[illegible]
	1889	[illegible]	[illegible]	[illegible]	[illegible]	[illegible]	[illegible]	[illegible]	[illegible]	[illegible]	[illegible]
CLARE,	1888	[illegible]	[illegible]	[illegible]	[illegible]	[illegible]	[illegible]	[illegible]	[illegible]	[illegible]	[illegible]
	1889	[illegible]	[illegible]	[illegible]	[illegible]	[illegible]	[illegible]	[illegible]	[illegible]	[illegible]	[illegible]
CORK,	1888	[illegible]	[illegible]	[illegible]	[illegible]	[illegible]	[illegible]	[illegible]	[illegible]	[illegible]	[illegible]
	1889	[illegible]	[illegible]	[illegible]	[illegible]	[illegible]	[illegible]	[illegible]	[illegible]	[illegible]	[illegible]
DONEGAL,	1888	[illegible]	[illegible]	[illegible]	[illegible]	[illegible]	[illegible]	[illegible]	[illegible]	[illegible]	[illegible]
	1889	[illegible]	[illegible]	[illegible]	[illegible]	[illegible]	[illegible]	[illegible]	[illegible]	[illegible]	[illegible]
DOWN,	1888	[illegible]	[illegible]	[illegible]	[illegible]	[illegible]	[illegible]	[illegible]	[illegible]	[illegible]	[illegible]
	1889	[illegible]	[illegible]	[illegible]	[illegible]	[illegible]	[illegible]	[illegible]	[illegible]	[illegible]	[illegible]
DUBLIN,	1888	[illegible]	[illegible]	[illegible]	[illegible]	[illegible]	[illegible]	[illegible]	[illegible]	[illegible]	[illegible]
	1889	[illegible]	[illegible]	[illegible]	[illegible]	[illegible]	[illegible]	[illegible]	[illegible]	[illegible]	[illegible]
FERMANAGH,	1888	[illegible]	[illegible]	[illegible]	[illegible]	[illegible]	[illegible]	[illegible]	[illegible]	[illegible]	[illegible]
	1889	[illegible]	[illegible]	[illegible]	[illegible]	[illegible]	[illegible]	[illegible]	[illegible]	[illegible]	[illegible]
GALWAY,	1888	[illegible]	[illegible]	[illegible]	[illegible]	[illegible]	[illegible]	[illegible]	[illegible]	[illegible]	[illegible]
	1889	[illegible]	[illegible]	[illegible]	[illegible]	[illegible]	[illegible]	[illegible]	[illegible]	[illegible]	[illegible]
KERRY,	1888	[illegible]	[illegible]	[illegible]	[illegible]	[illegible]	[illegible]	[illegible]	[illegible]	[illegible]	[illegible]
	1889	[illegible]	[illegible]	[illegible]	[illegible]	[illegible]	[illegible]	[illegible]	[illegible]	[illegible]	[illegible]
KILDARE,	1888	[illegible]	[illegible]	[illegible]	[illegible]	[illegible]	[illegible]	[illegible]	[illegible]	[illegible]	[illegible]
	1889	[illegible]	[illegible]	[illegible]	[illegible]	[illegible]	[illegible]	[illegible]	[illegible]	[illegible]	[illegible]
KILKENNY,	1888	[illegible]	[illegible]	[illegible]	[illegible]	[illegible]	[illegible]	[illegible]	[illegible]	[illegible]	[illegible]
	1889	[illegible]	[illegible]	[illegible]	[illegible]	[illegible]	[illegible]	[illegible]	[illegible]	[illegible]	[illegible]
KING'S,	1888	[illegible]	[illegible]	[illegible]	[illegible]	[illegible]	[illegible]	[illegible]	[illegible]	[illegible]	[illegible]
	1889	[illegible]	[illegible]	[illegible]	[illegible]	[illegible]	[illegible]	[illegible]	[illegible]	[illegible]	[illegible]
LEITRIM,	1888	[illegible]	[illegible]	[illegible]	[illegible]	[illegible]	[illegible]	[illegible]	[illegible]	[illegible]	[illegible]
	1889	[illegible]	[illegible]	[illegible]	[illegible]	[illegible]	[illegible]	[illegible]	[illegible]	[illegible]	[illegible]
LIMERICK,	1888	[illegible]	[illegible]	[illegible]	[illegible]	[illegible]	[illegible]	[illegible]	[illegible]	[illegible]	[illegible]
	1889	[illegible]	[illegible]	[illegible]	[illegible]	[illegible]	[illegible]	[illegible]	[illegible]	[illegible]	[illegible]
LONDONDERRY,	1888	[illegible]	[illegible]	[illegible]	[illegible]	[illegible]	[illegible]	[illegible]	[illegible]	[illegible]	[illegible]
	1889	[illegible]	[illegible]	[illegible]	[illegible]	[illegible]	[illegible]	[illegible]	[illegible]	[illegible]	[illegible]
LONGFORD,	1888	[illegible]	[illegible]	[illegible]	[illegible]	[illegible]	[illegible]	[illegible]	[illegible]	[illegible]	[illegible]
	1889	[illegible]	[illegible]	[illegible]	[illegible]	[illegible]	[illegible]	[illegible]	[illegible]	[illegible]	[illegible]
LOUTH & DROGHEDA Co. of TOWN,	1888	[illegible]	[illegible]	[illegible]	[illegible]	[illegible]	[illegible]	[illegible]	[illegible]	[illegible]	[illegible]
	1889	[illegible]	[illegible]	[illegible]	[illegible]	[illegible]	[illegible]	[illegible]	[illegible]	[illegible]	[illegible]
MAYO,	1888	[illegible]	[illegible]	[illegible]	[illegible]	[illegible]	[illegible]	[illegible]	[illegible]	[illegible]	[illegible]
	1889	[illegible]	[illegible]	[illegible]	[illegible]	[illegible]	[illegible]	[illegible]	[illegible]	[illegible]	[illegible]
MEATH,	1888	[illegible]	[illegible]	[illegible]	[illegible]	[illegible]	[illegible]	[illegible]	[illegible]	[illegible]	[illegible]
	1889	[illegible]	[illegible]	[illegible]	[illegible]	[illegible]	[illegible]	[illegible]	[illegible]	[illegible]	[illegible]
MONAGHAN,	1888	[illegible]	[illegible]	[illegible]	[illegible]	[illegible]	[illegible]	[illegible]	[illegible]	[illegible]	[illegible]
	1889	[illegible]	[illegible]	[illegible]	[illegible]	[illegible]	[illegible]	[illegible]	[illegible]	[illegible]	[illegible]
QUEEN'S,	1888	[illegible]	[illegible]	[illegible]	[illegible]	[illegible]	[illegible]	[illegible]	[illegible]	[illegible]	[illegible]
	1889	[illegible]	[illegible]	[illegible]	[illegible]	[illegible]	[illegible]	[illegible]	[illegible]	[illegible]	[illegible]
ROSCOMMON,	1888	[illegible]	[illegible]	[illegible]	[illegible]	[illegible]	[illegible]	[illegible]	[illegible]	[illegible]	[illegible]
	1889	[illegible]	[illegible]	[illegible]	[illegible]	[illegible]	[illegible]	[illegible]	[illegible]	[illegible]	[illegible]
SLIGO,	1888	[illegible]	[illegible]	[illegible]	[illegible]	[illegible]	[illegible]	[illegible]	[illegible]	[illegible]	[illegible]
	1889	[illegible]	[illegible]	[illegible]	[illegible]	[illegible]	[illegible]	[illegible]	[illegible]	[illegible]	[illegible]
TIPPERARY,	1888	[illegible]	[illegible]	[illegible]	[illegible]	[illegible]	[illegible]	[illegible]	[illegible]	[illegible]	[illegible]
	1889	[illegible]	[illegible]	[illegible]	[illegible]	[illegible]	[illegible]	[illegible]	[illegible]	[illegible]	[illegible]
TYRONE,	1888	[illegible]	[illegible]	[illegible]	[illegible]	[illegible]	[illegible]	[illegible]	[illegible]	[illegible]	[illegible]
	1889	[illegible]	[illegible]	[illegible]	[illegible]	[illegible]	[illegible]	[illegible]	[illegible]	[illegible]	[illegible]
WATERFORD,	1888	[illegible]	[illegible]	[illegible]	[illegible]	[illegible]	[illegible]	[illegible]	[illegible]	[illegible]	[illegible]
	1889	[illegible]	[illegible]	[illegible]	[illegible]	[illegible]	[illegible]	[illegible]	[illegible]	[illegible]	[illegible]
WESTMEATH,	1888	[illegible]	[illegible]	[illegible]	[illegible]	[illegible]	[illegible]	[illegible]	[illegible]	[illegible]	[illegible]
	1889	[illegible]	[illegible]	[illegible]	[illegible]	[illegible]	[illegible]	[illegible]	[illegible]	[illegible]	[illegible]
WEXFORD,	1888	[illegible]	[illegible]	[illegible]	[illegible]	[illegible]	[illegible]	[illegible]	[illegible]	[illegible]	[illegible]
	1889	[illegible]	[illegible]	[illegible]	[illegible]	[illegible]	[illegible]	[illegible]	[illegible]	[illegible]	[illegible]
WICKLOW,	1888	[illegible]	[illegible]	[illegible]	[illegible]	[illegible]	[illegible]	[illegible]	[illegible]	[illegible]	[illegible]
	1889	[illegible]	[illegible]	[illegible]	[illegible]	[illegible]	[illegible]	[illegible]	[illegible]	[illegible]	[illegible]

SUMMARY OF IRELAND.

PROVINCES.	Year	Not exceeding 1 Acre.	Above 1 and not exceeding 5 Acres.	Above 5 and not exceeding 15 Acres.	Above 15 and not exceeding 30 Acres.	Above 30 and not exceeding 50 Acres.	Above 50 and not exceeding 100 Acres.	Above 100 and not exceeding 200 Acres.	Above 200 and not exceeding 500 Acres.	Above 500 Acres.	Total.
LEINSTER,	1888	[illegible]	[illegible]	[illegible]	[illegible]	[illegible]	[illegible]	[illegible]	[illegible]	[illegible]	[illegible]
	1889	[illegible]	[illegible]	[illegible]	[illegible]	[illegible]	[illegible]	[illegible]	[illegible]	[illegible]	[illegible]
MUNSTER,	1888	[illegible]	[illegible]	[illegible]	[illegible]	[illegible]	[illegible]	[illegible]	[illegible]	[illegible]	[illegible]
	1889	[illegible]	[illegible]	[illegible]	[illegible]	[illegible]	[illegible]	[illegible]	[illegible]	[illegible]	[illegible]
ULSTER,	1888	[illegible]	[illegible]	[illegible]	[illegible]	[illegible]	[illegible]	[illegible]	[illegible]	[illegible]	[illegible]
	1889	[illegible]	[illegible]	[illegible]	[illegible]	[illegible]	[illegible]	[illegible]	[illegible]	[illegible]	[illegible]
CONNAUGHT,	1888	[illegible]	[illegible]	[illegible]	[illegible]	[illegible]	[illegible]	[illegible]	[illegible]	[illegible]	[illegible]
	1889	[illegible]	[illegible]	[illegible]	[illegible]	[illegible]	[illegible]	[illegible]	[illegible]	[illegible]	[illegible]
TOTAL OF IRELAND,	1888	[illegible]	[illegible]	[illegible]	[illegible]	[illegible]	[illegible]	[illegible]	[illegible]	[illegible]	[illegible]
	1889	[illegible]	[illegible]	[illegible]	[illegible]	[illegible]	[illegible]	[illegible]	[illegible]	[illegible]	[illegible]
INCREASE OR DECREASE in 1889,		Increase, [illegible]	Increase, [illegible]	Increase, [illegible]	Decrease, [illegible]	Decrease, [illegible]	Increase, [illegible]	Increase, [illegible]	Decrease, [illegible]	Increase, [illegible]	Increase, [illegible]

TABLE IV.—Return of the number of Occupiers resident in each County and Province in 1889, classified according to the total extent of land held, without reference to the Townland, Poor Law Union, County, or Province in which the portions of land are situated:—

Counties.	Number of Occupiers holding Land									
	Not exceeding 1 acre.	Above 1 and not exceeding 5 acres.	Above 5 and not exceeding 15 acres.	Above 15 and not exceeding 30 acres.	Above 30 and not exceeding 50 acres.	Above 50 and not exceeding 100 acres.	Above 100 and not exceeding 200 acres.	Above 200 and not exceeding 500 acres.	Above 500 acres.	Total.
Antrim,	[illegible]	[illegible]	[illegible]	[illegible]	[illegible]	[illegible]	[illegible]	[illegible]	[illegible]	[illegible]
Armagh,	[illegible]	[illegible]	[illegible]	[illegible]	[illegible]	[illegible]	[illegible]	[illegible]	[illegible]	[illegible]
Carlow,	[illegible]	[illegible]	[illegible]	[illegible]	[illegible]	[illegible]	[illegible]	[illegible]	[illegible]	[illegible]
Cavan,	[illegible]	[illegible]	[illegible]	[illegible]	[illegible]	[illegible]	[illegible]	[illegible]	[illegible]	[illegible]
Clare,	[illegible]	[illegible]	[illegible]	[illegible]	[illegible]	[illegible]	[illegible]	[illegible]	[illegible]	[illegible]
Cork,	[illegible]	[illegible]	[illegible]	[illegible]	[illegible]	[illegible]	[illegible]	[illegible]	[illegible]	[illegible]
Donegal,	[illegible]	[illegible]	[illegible]	[illegible]	[illegible]	[illegible]	[illegible]	[illegible]	[illegible]	[illegible]
Down,	[illegible]	[illegible]	[illegible]	[illegible]	[illegible]	[illegible]	[illegible]	[illegible]	[illegible]	[illegible]
Dublin,	[illegible]	[illegible]	[illegible]	[illegible]	[illegible]	[illegible]	[illegible]	[illegible]	[illegible]	[illegible]
Fermanagh,	[illegible]	[illegible]	[illegible]	[illegible]	[illegible]	[illegible]	[illegible]	[illegible]	[illegible]	[illegible]
Galway,	[illegible]	[illegible]	[illegible]	[illegible]	[illegible]	[illegible]	[illegible]	[illegible]	[illegible]	[illegible]
Kerry,	[illegible]	[illegible]	[illegible]	[illegible]	[illegible]	[illegible]	[illegible]	[illegible]	[illegible]	[illegible]
Kildare,	[illegible]	[illegible]	[illegible]	[illegible]	[illegible]	[illegible]	[illegible]	[illegible]	[illegible]	[illegible]
Kilkenny,	[illegible]	[illegible]	[illegible]	[illegible]	[illegible]	[illegible]	[illegible]	[illegible]	[illegible]	[illegible]
King's,	[illegible]	[illegible]	[illegible]	[illegible]	[illegible]	[illegible]	[illegible]	[illegible]	[illegible]	[illegible]
Leitrim,	[illegible]	[illegible]	[illegible]	[illegible]	[illegible]	[illegible]	[illegible]	[illegible]	[illegible]	[illegible]
Limerick,	[illegible]	[illegible]	[illegible]	[illegible]	[illegible]	[illegible]	[illegible]	[illegible]	[illegible]	[illegible]
Londonderry,	[illegible]	[illegible]	[illegible]	[illegible]	[illegible]	[illegible]	[illegible]	[illegible]	[illegible]	[illegible]
Longford,	[illegible]	[illegible]	[illegible]	[illegible]	[illegible]	[illegible]	[illegible]	[illegible]	[illegible]	[illegible]
[illegible] Town,	[illegible]	[illegible]	[illegible]	[illegible]	[illegible]	[illegible]	[illegible]	[illegible]	[illegible]	[illegible]
Mayo,	[illegible]	[illegible]	[illegible]	[illegible]	[illegible]	[illegible]	[illegible]	[illegible]	[illegible]	[illegible]
Meath,	[illegible]	[illegible]	[illegible]	[illegible]	[illegible]	[illegible]	[illegible]	[illegible]	[illegible]	[illegible]
Monaghan,	[illegible]	[illegible]	[illegible]	[illegible]	[illegible]	[illegible]	[illegible]	[illegible]	[illegible]	[illegible]
Queen's,	[illegible]	[illegible]	[illegible]	[illegible]	[illegible]	[illegible]	[illegible]	[illegible]	[illegible]	[illegible]
Roscommon,	[illegible]	[illegible]	[illegible]	[illegible]	[illegible]	[illegible]	[illegible]	[illegible]	[illegible]	[illegible]
Sligo,	[illegible]	[illegible]	[illegible]	[illegible]	[illegible]	[illegible]	[illegible]	[illegible]	[illegible]	[illegible]
Tipperary,	[illegible]	[illegible]	[illegible]	[illegible]	[illegible]	[illegible]	[illegible]	[illegible]	[illegible]	[illegible]
Tyrone,	[illegible]	[illegible]	[illegible]	[illegible]	[illegible]	[illegible]	[illegible]	[illegible]	[illegible]	[illegible]
Waterford,	[illegible]	[illegible]	[illegible]	[illegible]	[illegible]	[illegible]	[illegible]	[illegible]	[illegible]	[illegible]
Westmeath,	[illegible]	[illegible]	[illegible]	[illegible]	[illegible]	[illegible]	[illegible]	[illegible]	[illegible]	[illegible]
Wexford,	[illegible]	[illegible]	[illegible]	[illegible]	[illegible]	[illegible]	[illegible]	[illegible]	[illegible]	[illegible]
Wicklow,	[illegible]	[illegible]	[illegible]	[illegible]	[illegible]	[illegible]	[illegible]	[illegible]	[illegible]	[illegible]
SUMMARY OF IRELAND.										
PROVINCES.										
Leinster,	[illegible]	[illegible]	[illegible]	[illegible]	[illegible]	[illegible]	[illegible]	[illegible]	[illegible]	[illegible]
Munster,	[illegible]	[illegible]	[illegible]	[illegible]	[illegible]	[illegible]	[illegible]	[illegible]	[illegible]	[illegible]
Ulster,	[illegible]	[illegible]	[illegible]	[illegible]	[illegible]	[illegible]	[illegible]	[illegible]	[illegible]	[illegible]
Connaught,	[illegible]	[illegible]	[illegible]	[illegible]	[illegible]	[illegible]	[illegible]	[illegible]	[illegible]	[illegible]
Total of Ireland,	[illegible]	[illegible]	[illegible]	[illegible]	[illegible]	[illegible]	[illegible]	[illegible]	[illegible]	[illegible]

The following statement shows the **number of occupiers of land in** each year from 1883 to 1889, by Provinces:— Number of Occupiers of Land, 1883 to 1889.

Provinces.	Number of Occupiers in the Years.						
	[illegible]	[illegible]	[illegible]	[illegible]	[illegible]	[illegible]	[illegible]
Leinster,	[illegible]	[illegible]	[illegible]	[illegible]	[illegible]	[illegible]	[illegible]
Munster,	[illegible]	[illegible]	[illegible]	[illegible]	[illegible]	[illegible]	[illegible]
Ulster,	[illegible]	[illegible]	[illegible]	[illegible]	[illegible]	[illegible]	[illegible]
Connaught,	[illegible]	[illegible]	[illegible]	[illegible]	[illegible]	[illegible]	[illegible]
Ireland,	[illegible]	[illegible]	[illegible]	[illegible]	[illegible]	[illegible]	[illegible]

The number of holdings "above 1 and not exceeding 5 acres" diminished greatly between 1841 and 1889. In Leinster the decrease was 65·1 per cent.; in Munster 81·2; in Ulster 79·6; in Connaught 87·6; and in all Ireland 80·2 per cent. Increase or decrease in Holdings by Classes between 1841 and 1889.

In the same period holdings "above 5 and not exceeding 15 acres" also diminished in number; the decrease in all Ireland was 38·1 per cent.; it was—in Leinster 44·8 per cent.; in Munster 69·6; and in Ulster 38·0; while in Connaught these holdings increased 2·4 per cent.

Holdings "above 15 and not exceeding 30 acres" increased 7·5 per cent. in Leinster; 116·3 per cent. in Ulster; and 489·8 per cent. in Connaught. They decreased 12·0 per cent. in Munster; while in all Ireland they increased 70·3 per cent.

Holdings "above 30 acres" increased 119·5 per cent. in Leinster; 240·4 in Munster; 353·0 in Ulster; 426·2 in Connaught; and 234·8 per cent. in all Ireland.

The total number of holdings "above 1 acre" decreased between 1841 and 1889 by 22·4 per cent. in Leinster; 32·5 per cent. in Munster; 21·8 in Ulster; and 25·7 in Connaught.

The total number of holdings in Ireland "above 1 acre" was 691,202 in 1841; 570,338 in 1851; 568,484 in 1861; 544,148 in 1871; 528,748 in 1881; and 516,046 in 1889, showing a decrease of 175,156 or 25·3 per cent. in the period between 1841 and 1889.

Number of Holdings in 1841, 1851, 1861, 1871, 1881, and 1889.

TABLE V.—The number of Holdings above 1 acre in each Province in 1841, 1851, 1861, 1871, 1881, and 1889, according to the classification used by the Census Commissioners of 1841 (in which "above 30 acres" was the maximum); the increase or decrease in the numbers in each class, and the differences per cent., between 1841 and 1889:—

Size of Holdings.		Leinster.	Munster.	Ulster.	Connaught.	Ireland.
		Number.	Number.	Number.	Number.	Number.
Above 1 and not exceeding 5 Acres,	1841,	54,119	57,857	100,816	100,254	310,436
	1851,	15,214	14,380	39,709	18,448	88,083
	1861,	23,848	23,738	24,658	14,517	85,469
	1871,	21,189	15,208	24,202	10,850	74,809
	1881,	16,864	11,094	21,871	15,809	62,071
	1889,	17,401	10,873	20,819	12,938	61,530
		Decrease.	Decrease.	Decrease.	Decrease.	Decrease.
Decrease in number between 1841 and 1889,		36,668	46,980	81,303	87,196	248,918
Rate per cent.,		80·1	81·2	79·9	87·9	80·2
Above 5 and not exceeding 15 Acres,	1841,	68,030	62,728	93,605	41,402	252,799
	1851,	34,000	24,100	86,116	42,280	191,854
	1861,	29,816	31,918	82,035	60,534	183,931
	1871,	27,376	31,400	73,047	50,102	171,383
	1881,	30,048	19,317	63,318	46,888	164,045
	1889,	25,423	14,773	64,873	41,[illegible]	158,401
		Decrease.	Decrease.	Decrease.	Increase.	Decrease.
Increase or Decrease in number between 1841 and 1889,		36,816	43,378	28,732	1,028	94,398
Rate per cent.,		[illegible]	69·4	30·0	[illegible]	36·1
Above 15 and not exceeding 30 Acres,	1841,	20,050	27,611	25,518	6,914	79,342
	1851,	26,630	38,328	47,541	31,797	141,311
	1861,	24,825	37,404	47,040	34,500	141,251
	1871,	23,348	37,214	46,278	32,729	138,647
	1881,	24,953	[illegible]	46,227	32,913	133,793
	1889,	22,810	24,967	44,868	32,044	130,096
		Increase.	Decrease.	Increase.	Increase.	Increase.
Increase or Decrease in number between 1841 and 1889,		1,472	2,684	19,340	26,130	50,754
Rate per cent.,		7·3	10·9	115·3	409·8	76·3
Above 30 Acres,	1841,	17,068	15,666	9,846	4,361	48,625
	1851,	52,828	39,224	37,812	20,767	149,090
	1861,	53,389	55,833	39,484	22,157	157,833
	1871,	56,431	55,498	41,071	22,273	159,303
	1881,	35,473	56,141	41,510	31,798	157,831
	1889,	38,878	56,280	43,941	22,024	162,790
		Increase.	Increase.	Increase.	Increase.	Increase.
Increase in number between 1841 and 1889,		21,400	40,681	34,095	18,528	114,178
Rate per cent.,		119·5	240·4	353·0	426·2	234·8
Total above 1 Acre,	1841,	134,790	162,856	236,694	155,842	691,202
	1851,	120,071	126,424	211,349	114,834	570,338
	1861,	116,073	118,893	193,888	139,643	568,484
	1871,	111,478	111,742	185,325	121,863	544,149
	1881,	106,960	112,614	169,056	112,708	528,748
	1889,	104,618	110,680	184,995	110,773	516,046
		Decrease.	Decrease.	Decrease.	Decrease.	Decrease.
Decrease in number between 1841 and 1889,		30,162	52,256	51,698	40,070	175,156
Rate per cent.,		22·4	32·5	21·8	25·7	25·3

PART II.—THE PRODUCE OF THE CROPS.

The Tables relating to the *produce* of the crops have been carefully compiled from information obtained by members of the Royal Irish Constabulary and of the Metropolitan Police from practical farmers and other persons qualified to form an opinion as to the yield in that *Poor Law Electoral Division* (adopted since 1855, instead of Constabulary Districts), for which they were requested to afford the information. The names and residences of the parties so co-operating and assisting are stated by the Enumerators on the Returns. Mode of collecting the Returns of Produce.

Conditions influencing the Produce of the Crops.

Notes of Superintendents of Enumeration.

On pp. 60 to 69 will be found the observations of the District Inspectors of the Royal Irish Constabulary and of the Sergeants of the Metropolitan Police, who acted as Superintendents of Enumeration, in reply to a circular requesting their opinion on the probable cause to which the good or bad yield of the various crops, in each of their districts, may be attributed.

The Weather.

The Weather being a potent factor in influencing the produce of the crops, both as to quantity and quality, the following particulars and those given on pages 123–144 are inserted by the kind permission of the Editor of the Dublin Journal of Medical Science: they have been derived from Returns of Meteorological Observations taken in Dublin City during the years 1869–88, by J. W. Moore, Esq., M.D., F.K.Q.C.P.I., F.R. MET. SOC.; and published in the Journal during the years 1889–90. The Tables on pages 145–7 also, are founded on Dr. Moore's observations:— The Weather.

The mean Atmospherical Pressure has been obtained from daily readings of the barometer at 9 A.M. and 9 P.M., corrected and reduced to 32° Fahrenheit at the mean sea level. The Mean Temperature values have been deduced from the maximal and minimal readings of the thermometer in the shade by Kaemtz's Formula, viz., min. + (max. – min. × ·41) = Mean Temperature. The Rainfall is that measured daily at 9 A.M. A rainy day is one on which at least one-hundredth (·01) of an inch of rain falls within the twenty-four hours from 9 A.M. to 9 A.M.

The *Mean Height of the Barometer* during the year 1889 was 29·908 inches. The highest observed reading was 30·780 inches at 9 P.M. on December 6th. The lowest observed reading was 28·740 inches at 4 A.M. on October 7th. The extreme range of atmospherical pressure was 1·940 inches compared with 2·130 inches in 1888.

The *Mean Temperature* of the year, deduced from the maximal and minimal readings of the thermometer in the shade by Kaemtz's Formula, was 48·4°. The highest reading was 77·8° on July 8th; the lowest reading was 21·7° on February 11th. The average mean temperature for the years 1865–89 calculated in the same way, was 48·6°. The mean temperature deduced from the daily readings of the dry bulb thermometer at 9 A.M. and 9 P.M. was 49·2°.

Rain fell on 198 days, including snow or sleet on 15 days, and hail on 20 days. The average number of rainy days in the years 1869–88 was 190·8. The total rainfall measured 27·972 inches, compared with an average of 27·382 inches in the twenty years 1865–84. During the first half of 1889 (January to June, inclusive) the rainfall was 10·876 inches on 92 days; during the second half (July to December, inclusive) 16·096 inches fell on 96 days.

As regards the *Direction of the Wind*, 730 observations were made during the year, with this result—N., 65; N.E., 40; E., 35; S.E., 56; S., 73; S.W., 98; W., 181; N.W., 112; Calms, 42.

Noxious Insects.

The ravages of noxious insects are frequently mentioned in the notes of the Superintendents of Enumeration. These insects have hitherto attracted but little notice in Ireland as compared with Great Britain and Foreign Countries. Mr. Mathewson, the Secretary of this Department, has given a great deal of attention to the natural history of insect pests and the injuries inflicted on crops by them, and has kindly placed at my disposal some valuable notes which he has compiled with regard to this subject, accompanied by illustrative drawings. I propose as soon as possible to publish these notes with the view of supplying the Superintendents of Enumeration and others interested in the question with accurate information regarding noxious insects. Mr. Mathewson received the kind advice and assistance of several distinguished entomologists when preparing his interesting paper which will be issued as a supplement to this Report together with some notes on noxious weeds. Noxious Insects.

Weeds and Noxious Plants.

Weeds and Noxious Plants.

For many years it was the custom of the late Mr. Donnelly, when Registrar-General, to point out the great injury done to crops by the very plentiful distribution of weeds in Ireland. Weeds not only interfere with the growth of the crops with which they are mingled but also exhaust the ground; in some cases are injurious to cattle, and even, though in rare instances, produce seeds which when mixed with grain produce poisonous effects on the consumers of food made from such grain. Besides weeds, properly so called, there are several injurious fungi which damage crops, fruits, trees, and vegetables of various kinds.

Mr. Donnelly issued circulars at various times warning farmers and others against the injurious effects of weeds. It is probable that these circulars would have had better effect if accompanied by more detailed information concerning the pests which they dealt with. I propose to endeavour to remedy this defect by issuing with the supplement on noxious insects some detailed information on noxious weeds. As in the case of the insects, Mr. Matheson and I have been much aided in arranging the information on the subject by scientific friends.

Total produce in 1888 and 1889.

Comparing the total produce of the crops in 1889 with the total in 1888—the returns for 1889 are, with the exception of flax and hay, of a satisfactory character. In Cereal Crops, there is an increase in wheat of 68,224 cwts.; in oats of 2,007 cwts.; in barley of 340,853 cwts.; in bere of 1,096 cwts.; in rye of 16,283 cwts.; and in beans of 3,271; there is a decrease in pease of 530 cwts.

In Green Crops, there is an increase in potatoes of 334,415 tons in 1889 compared with 1888, an increase of 583,171 tons in turnips, of 31,584 tons in mangel wurzel and beet root, and of 45,177 tons in cabbage.

Flax shows a decrease of 197,179 stones of 14 lbs.; hay on clover, sainfoin, and grasses under rotation, an increase of 91,586 tons; while hay on permanent pasture or grass not broken up in rotation, shows a decrease of 418,531 tons.

Estimated average produce per acre in 1888 and 1889.

The yield per acre of cereal crops in 1889 compared with that of 1888 shows an increase in wheat of 2·2 cwts.; in oats of 0·4 cwt.; in barley of 1·7 cwt.; in beans of 5·6 cwts.; and in pease of 0·1 cwt., while there is a decrease in bere of 0·3 cwt., and in rye of 0·5 cwt. In other crops—potatoes show an increase of 0·5 ton; turnips of 1·8 tons; mangel wurzel and beet root of 1·2 tons; and cabbage of 1·1 tons. Hay on clover, sainfoin, and grasses under rotation shows the same rate in both years, and hay on permanent pasture or grass not broken up in rotation, decreased by 0·2 ton. Flax shows a decrease of 1·7 stones.

The total *produce* of the principal crops in 1888 and 1889, and the increase or decrease in the latter year, are given in Table VI.; the *average produce* per statute acre in Table VII.; and in Table VIII. are given the total extent under each of the principal crops, the *estimated average* yield per statute acre, and the total produce, for each year from 1880 to 1889, inclusive.

Produce of the Crops, 1888–89.

TABLE VI.—The *total produce* of the principal Crops in 1888 and 1889, and the increase or decrease in the latter year:—

Crops.	Produce. 1888.	Produce. 1889.	Increase in 1889.	Decrease in 1889.
Wheat, Cwts. of 112 lbs.,	1,367,029	1,435,[illegible]	68,224	—
Oats, " "	17,356,538	17,358,545	2,007	—
Barley, " "	3,501,829	3,842,682	340,853	—
Bere, " "	6,043	8,139	1,096	—
Rye, " "	187,[illegible]	203,435	16,283	—
Beans, " "	63,[illegible]	67,270	3,271	—
Pease, " "	5,269	4,739	—	530
Potatoes, in Tons,	2,[illegible]	2,847,632	334,415	—
Turnips, "	3,323,651	3,906,822	583,171	—
Mangel Wurzel and Beet Root, "	802,005	831,689	31,584	—
Cabbage, "	385,445	431,[illegible]	45,177	—
Flax, in Stones of 14 lbs.,	2,255,720	2,058,541	—	197,179
Hay, in Tons. Clover, Sainfoin, and Grasses under Rotation,	1,389,758	1,478,344	91,586	—
Hay, in Tons. Permanent Pasture or Grass not broken up in Rotation,	3,794,138	3,375,607	—	418,531

TABLE VII.—The estimated average produce per statute acre of the principal crops in 1888 and 1889, and the increase or decrease in 1889 compared with 1888:— Average produce of Crops in 1888 and 1889.

Crops.	Produce per Statute Acre. 1888.	1889.	Increase in 1889.	Decrease in 1889.
Wheat, in Cwts. of 112 lbs.,	13·8	16·0	2·2	—
Oats, " "	13·8	14·2	0·4	—
Barley, " "	15·8	17·5	1·7	—
Bere, " "	13·3	13·0	—	0·3
Rye, " "	13·4	12·9	—	0·5
Beans, " "	12·0	13·1	5·6	—
Peas, " "	11·8	15·4	3·4	—
Potatoes, in Tons,	3·1	3·4	0·3	—
Turnips, "	13·8	15·4	1·6	—
Mangel Wurzel and Beet Root, "	12·9	14·1	1·2	—
Cabbage, "	9·1	10·2	1·1	—
Flax, in Stones of 14 lbs.,	23·0	21·3	—	1·7
Hay, in Tons: Clover, Sainfoin, and Grasses under Rotation,	2·2	2·3	—	—
Hay, in Tons: Permanent Pasture or Grass not broken up in Rotation,	2·4	2·2	—	0·2

The further statement contained in Table VIII. gives a general view of the state of agriculture during the year 1889 as compared with preceding years. Extent under Crops, produce, &c., 1880–89.

Tables showing the total produce of the Crops in 1889, by counties and provinces, will be found at page 26, and by poor law unions at page 38. The average rates by counties and provinces for each year from 1880 to 1889, are given at pages 41 to 45.

TABLE VIII.—The extent under each of the principal Crops—the average Yield per Statute Acre, and the total Produce for all Ireland, in each year from 1880 to 1889, inclusive.

Years.	Wheat.	Oats.	Barley.	Bere.	Rye.	Potatoes.	Turnips.	Mangel Wurzel and Beet Root.	Cabbage.	Flax.	Hay.
	EXTENT UNDER CROPS IN STATUTE MEASURE.										
	Acres.	Acres.	Acres.	Acres.	Acres.	Acres.	Acres.	Acres.	Acres.	Acres.	Acres.
1880,	[illegible]	[illegible]	[illegible]	[illegible]	[illegible]	[illegible]	[illegible]	[illegible]	[illegible]	[illegible]	[illegible]
1881,	[illegible]	[illegible]	[illegible]	[illegible]	[illegible]	[illegible]	[illegible]	[illegible]	[illegible]	[illegible]	[illegible]
1882,	[illegible]	[illegible]	[illegible]	[illegible]	[illegible]	[illegible]	[illegible]	[illegible]	[illegible]	[illegible]	[illegible]
1883,	[illegible]	[illegible]	[illegible]	[illegible]	[illegible]	[illegible]	[illegible]	[illegible]	[illegible]	[illegible]	[illegible]
1884,	[illegible]	[illegible]	[illegible]	[illegible]	[illegible]	[illegible]	[illegible]	[illegible]	[illegible]	[illegible]	[illegible]
1885,	[illegible]	[illegible]	[illegible]	[illegible]	[illegible]	[illegible]	[illegible]	[illegible]	[illegible]	[illegible]	[illegible]
1886,	[illegible]	[illegible]	[illegible]	[illegible]	[illegible]	[illegible]	[illegible]	[illegible]	[illegible]	[illegible]	[illegible]
1887,	[illegible]	[illegible]	[illegible]	[illegible]	[illegible]	[illegible]	[illegible]	[illegible]	[illegible]	[illegible]	[illegible]
1888,	[illegible]	[illegible]	[illegible]	[illegible]	[illegible]	[illegible]	[illegible]	[illegible]	[illegible]	[illegible]	[illegible]
1889,	[illegible]	[illegible]	[illegible]	[illegible]	[illegible]	[illegible]	[illegible]	[illegible]	[illegible]	[illegible]	[illegible]
—	ESTIMATED AVERAGE PRODUCE PER STATUTE ACRE.										
	Cwts. of 112 lbs.	Cwts. of 112 lbs.	Cwts. of 112 lbs.	Cwts. of 112 lbs.	Cwts. of 112 lbs.	Tons.	Tons.	Tons.	Tons.	Stones of 14 lbs.	Tons.
1880,	[illegible]	[illegible]	[illegible]	[illegible]	[illegible]	[illegible]	[illegible]	[illegible]	[illegible]	[illegible]	[illegible]
1881,	[illegible]	[illegible]	[illegible]	[illegible]	[illegible]	[illegible]	[illegible]	[illegible]	[illegible]	[illegible]	[illegible]
1882,	[illegible]	[illegible]	[illegible]	[illegible]	[illegible]	[illegible]	[illegible]	[illegible]	[illegible]	[illegible]	[illegible]
1883,	[illegible]	[illegible]	[illegible]	[illegible]	[illegible]	[illegible]	[illegible]	[illegible]	[illegible]	[illegible]	[illegible]
1884,	[illegible]	[illegible]	[illegible]	[illegible]	[illegible]	[illegible]	[illegible]	[illegible]	[illegible]	[illegible]	[illegible]
1885,	[illegible]	[illegible]	[illegible]	[illegible]	[illegible]	[illegible]	[illegible]	[illegible]	[illegible]	[illegible]	[illegible]
1886,	[illegible]	[illegible]	[illegible]	[illegible]	[illegible]	[illegible]	[illegible]	[illegible]	[illegible]	[illegible]	[illegible]
1887,	[illegible]	[illegible]	[illegible]	[illegible]	[illegible]	[illegible]	[illegible]	[illegible]	[illegible]	[illegible]	[illegible]
1888,	[illegible]	[illegible]	[illegible]	[illegible]	[illegible]	[illegible]	[illegible]	[illegible]	[illegible]	[illegible]	[illegible]
1889,	[illegible]	[illegible]	[illegible]	[illegible]	[illegible]	[illegible]	[illegible]	[illegible]	[illegible]	[illegible]	[illegible]
—	TOTAL PRODUCE.										
	Cwts. of 112 lbs.	Cwts. of 112 lbs.	Cwts. of 112 lbs.	Cwts. of 112 lbs.	Cwts. of 112 lbs.	Tons.	Tons.	Tons.	Tons.	Stones.	Tons.
1880,	[illegible]	[illegible]	[illegible]	[illegible]	[illegible]	[illegible]	[illegible]	[illegible]	[illegible]	[illegible]	[illegible]
1881,	[illegible]	[illegible]	[illegible]	[illegible]	[illegible]	[illegible]	[illegible]	[illegible]	[illegible]	[illegible]	[illegible]
1882,	[illegible]	[illegible]	[illegible]	[illegible]	[illegible]	[illegible]	[illegible]	[illegible]	[illegible]	[illegible]	[illegible]
1883,	[illegible]	[illegible]	[illegible]	[illegible]	[illegible]	[illegible]	[illegible]	[illegible]	[illegible]	[illegible]	[illegible]
1884,	[illegible]	[illegible]	[illegible]	[illegible]	[illegible]	[illegible]	[illegible]	[illegible]	[illegible]	[illegible]	[illegible]
1885,	[illegible]	[illegible]	[illegible]	[illegible]	[illegible]	[illegible]	[illegible]	[illegible]	[illegible]	[illegible]	[illegible]
1886,	[illegible]	[illegible]	[illegible]	[illegible]	[illegible]	[illegible]	[illegible]	[illegible]	[illegible]	[illegible]	[illegible]
1887,	[illegible]	[illegible]	[illegible]	[illegible]	[illegible]	[illegible]	[illegible]	[illegible]	[illegible]	[illegible]	[illegible]
1888,	[illegible]	[illegible]	[illegible]	[illegible]	[illegible]	[illegible]	[illegible]	[illegible]	[illegible]	[illegible]	[illegible]
1889,	[illegible]	[illegible]	[illegible]	[illegible]	[illegible]	[illegible]	[illegible]	[illegible]	[illegible]	[illegible]	[illegible]

PART III.—LIVE STOCK.

Number and Ages of Live Stock, 1888 and 1889.

TABLE IX.—The Number and Ages of the Live Stock in Ireland, in 1888 and 1889, and the Increase or Decrease in each description:—

Description of Stock.		Number in 1888.	Number in 1889.	Increase in 1889.	Decrease in 1889.
HORSES,	Two years old and upwards,	470,341	477,307	6,966	—
	One year old and under two,	74,263	77,062	2,799	—
	Under one year,	70,483	69,895	—	588
Total No. of Horses,		615,087	624,264	9,167	—
MULES,		30,271	29,838	—	433
ASSES,		203,152	206,236	3,084	—
CATTLE,	Two years old and upwards,	2,303,529	2,274,038	—	29,491
	One year old and under two,	871,045	863,835	—	7,210
	Under one year,	901,521	983,201	81,680	—
Total No. of Cattle,		4,099,205	4,094,174	—	5,031
SHEEP,	One year old and upwards,	2,151,116	2,262,846	111,730	—
	Under one year,	1,455,553	1,526,341	70,788	—
Total No. of Sheep,		3,606,669	3,789,187	182,518	—
PIGS,	One year old and upwards,	171,091	168,645	—	2,446
	Under one year,	1,226,735	1,212,025	—	14,710
Total No. of Pigs,		1,397,826	1,380,670	—	17,156
GOATS,		295,378	303,633	8,255	—
POULTRY,		14,486,400	14,856,517	370,117	—

Number of Live Stock.

At the period of the enumeration in 1889, the total number of horses in Ireland was 574,264, being an *increase* of 9,167 compared with 1888. There was an *increase* of 6,966 in the number "two years old and upwards," and of 2,799 in the "one year old and under two," while there was a *decrease* of 588 in those "under one year."

Mules numbered 29,838, being 433 less than in 1888, and asses 206,236, being an *increase* of 3,084 as compared with the previous year.

Horses, Mules and Asses taken together numbered 768,457 in 1880, and 810,338 in 1889, being an *increase* of 41,881 or 5·5 per cent.

Cattle numbered 4,094,174 in 1889, showing a total *decrease* of 5,031 as compared with the number enumerated in 1888; in the "two years old and upwards" there was a *decrease* of 29,491, and in the "one year old and under two" a *decrease* of 7,210; while there was an *increase* of 81,680 in the number "under one year." Taking the ten years 1880 to 1889, cattle *increased* in number from 3,921,317 in 1880, to 4,228,851 in 1885, but *decreased* in each of the four following years, the number for 1889 being 4,094,174 as already stated.

Sheep amounted to 3,789,187 in 1889, showing an *increase* of 162,516, as compared with the previous year: the "one year old and upwards" *increased* by 91,730, and those "under one year" by 70,786. Number of Live Stock.

Comparing 1880 with 1889 there has been an *increase* in the number of sheep from 3,562,463 in the former, to 3,789,187 in the latter year.

Pigs were returned as 1,380,670 in 1889, showing a *decrease* of 17,155, or 1·2 per cent. as compared with the previous year. The "one year old and upwards" *decreased* by 2,445, and those "under one year" by 14,710.

Comparing the number of pigs returned in the ten years from 1880 to 1889, the highest number, 1,430,128, was enumerated in 1882, and the lowest, 850,269, in 1880.

Goats numbered 303,923 in 1889, being 8,245 more than in 1888, and 38,144 more than in 1880.

The number of poultry in 1889 was 14,856,517, being 370,117 more than in 1888, and 1,426,335 more than in 1880. Of the 14,856,517 poultry in 1889, 985,456 were turkeys; 1,150,581 geese; 2,911,737 ducks; and 6,808,743 ordinary fowl. Poultry.

Compared with 1888, turkeys *increased* by 39,402, geese by 38,221, ducks by 49,993, and ordinary fowl by 282,498.

TABLE X.—The Number of *Live Stock* in Ireland, in each year from 1880 to 1889, inclusive:— Number of Live Stock, 1880 to 1889.

Years.	Horses and Mules.	Asses.	Cattle.	Sheep.	Pigs.	Goats.	Poultry.
1880, . .	552,130	166,387	3,921,517	3,562,463	850,269	265,739	13,430,182
1881, . .	574,746	167,143	3,956,595	3,256,188	1,096,830	260,078	13,972,438
1882, . .	555,958	167,763	3,987,811	3,071,758	1,430,128	253,373	13,399,086
1883, . .	541,427	169,760	4,096,858	3,210,611	1,348,364	263,148	13,352,430
1884, . .	549,436	191,839	4,112,760	3,245,219	1,306,350	254,411	12,747,460
1885, . .	570,429	197,170	4,228,851	3,478,006	1,269,083	264,437	13,890,932
1886, . .	578,259	196,345	4,183,851	3,364,043	1,263,143	264,178	13,509,822
1887, . .	547,234	180,513	4,157,404	3,377,936	1,408,658	271,739	14,460,643
1888, . .	536,348	203,153	4,098,188	3,626,669	1,397,825	295,678	14,486,400
1889, . .	594,109	205,336	4,091,174	3,789,187	1,380,670	303,923	14,856,517

TABLE XI.—The proportion per cent. of Horses, Cattle, Sheep, and Pigs in Ireland, according to Age, for the years 1880 to 1889, inclusive:— Number of Live Stock, 1880 to 1889.

Years.	Horses. Percentage at each age.			Cattle. Percentage at each age.			Sheep. Percentage at each age.		Pigs. Percentage at each age.	
	Two Years old and upwards.	One Year old and under Two.	Under One Year.	Two Years old and upwards.	One Year old and under Two.	Under One Year.	One Year old and upwards.	Under One Year.	One Year old and upwards.	Under One Year.
1880, .	78·2	12·1	9·7	57·7	20·9	21·4	64·4	35·6	13·9	86·1
1881, .	78·2	11·4	9·4	57·9	19·9	22·2	64·5	35·5	13·7	86·3
1882, .	79·6	10·4	10·0	57·0	17·7	25·1	63·0	37·0	13·9	86·6
1883, .	79·3	10·6	10·3	56·3	20·6	23·9	61·7	38·3	13·4	86·6
1884, .	78·0	11·1	10·9	55·6	21·6	23·2	62·5	37·5	12·8	87·2
1885, .	76·6	11·6	11·8	56·9	20·8	22·2	61·5	38·5	12·7	87·3
1886, .	76·3	12·3	11·4	56·7	21·0	22·3	61·7	38·3	12·7	87·3
1887, .	75·6	12·6	11·7	56·7	20·8	22·5	60·4	39·6	12·7	87·3
1888, .	74·4	13·1	12·5	56·9	21·6	21·5	59·6	40·4	12·5	87·5
1889, .	74·4	13·4	12·2	55·6	21·5	22·9	59·5	40·5	12·2	87·8

MILCH COWS.

TABLE XII.—The following statement shows the number of Milch Cows in Ireland in each year from 1854—the first year in which Milch Cows were separately enumerated—to 1889. The average number for the first five years of the period was 1,579,851, and for the last five years 1,395,751 being a decline of 184,100 or 11·7 per cent. The highest number in any one year was 1,690,389 in 1859, and the lowest 1,348,986 in 1864.

Years.	No. of Milch Cows.	Years.	No. of Milch Cows.	Years.	No. of Milch Cows.	Years.	No. of Milch Cows.
1854,	1,517,673	1863,	[illegible]	1872,	[illegible]	1881,	[illegible]
1855,	[illegible]	1864,	1,348,986	1873,	[illegible]	1882,	[illegible]
1856,	[illegible]	1865,	[illegible]	1874,	[illegible]	1883,	[illegible]
1857,	[illegible]	1866,	[illegible]	1875,	[illegible]	1884,	[illegible]
1858,	[illegible]	1867,	[illegible]	1876,	[illegible]	1885,	[illegible]
1859,	[illegible]	1868,	[illegible]	1877,	[illegible]	1886,	[illegible]
1860,	[illegible]	1869,	[illegible]	1878,	[illegible]	1887,	[illegible]
1861,	[illegible]	1870,	[illegible]	1879,	[illegible]	1888,	[illegible]
1862,	[illegible]	1871,	[illegible]	1880,	[illegible]	1889,	[illegible]

Tables showing the number of Live Stock in 1889, by counties and provinces will be found at page 46; by Poor Law Unions at pages 47-50; and by counties and provinces for each year from 1880 to 1889 at page 51.

EXPORTS AND IMPORTS OF LIVE STOCK.

With the view of giving a more accurate idea of the number of live stock produced in Ireland the following statement has been extracted from the Statistical Returns published in the Report for 1889 under the "Contagious Diseases (Animals) Act, 1878."

Number of Cattle, Sheep, and Swine, exported from Ireland to Great Britain during each of the fifteen years 1875–89 :—

Years.	Cattle.						Sheep.			Swine.			Years.
	Oxen, Bulls, and Cows.				Calves.	Total.	Sheep.	Lambs.	Total.	Fat Swine.	Store Swine.	Total.	
	Fat Cattle.	Store Cattle for Grazing or Breeding purposes.	Other Cattle.	Total.									
[illegible]	[illegible]	[illegible]	[illegible]	[illegible]	[illegible]	[illegible]	[illegible]	[illegible]	[illegible]	[illegible]	[illegible]	[illegible]	[illegible]
[illegible]	[illegible]	[illegible]	[illegible]	[illegible]	[illegible]	[illegible]	[illegible]	[illegible]	[illegible]	[illegible]	[illegible]	[illegible]	[illegible]
[illegible]	[illegible]	[illegible]	[illegible]	[illegible]	[illegible]	[illegible]	[illegible]	[illegible]	[illegible]	[illegible]	[illegible]	[illegible]	[illegible]
[illegible]	[illegible]	[illegible]	[illegible]	[illegible]	[illegible]	[illegible]	[illegible]	[illegible]	[illegible]	[illegible]	[illegible]	[illegible]	[illegible]
[illegible]	[illegible]	[illegible]	[illegible]	[illegible]	[illegible]	[illegible]	[illegible]	[illegible]	[illegible]	[illegible]	[illegible]	[illegible]	[illegible]
[illegible]	[illegible]	[illegible]	[illegible]	[illegible]	[illegible]	[illegible]	[illegible]	[illegible]	[illegible]	[illegible]	[illegible]	[illegible]	[illegible]
[illegible]	[illegible]	[illegible]	[illegible]	[illegible]	[illegible]	[illegible]	[illegible]	[illegible]	[illegible]	[illegible]	[illegible]	[illegible]	[illegible]
[illegible]	[illegible]	[illegible]	[illegible]	[illegible]	[illegible]	[illegible]	[illegible]	[illegible]	[illegible]	[illegible]	[illegible]	[illegible]	[illegible]
[illegible]	[illegible]	[illegible]	[illegible]	[illegible]	[illegible]	[illegible]	[illegible]	[illegible]	[illegible]	[illegible]	[illegible]	[illegible]	[illegible]
[illegible]	[illegible]	[illegible]	[illegible]	[illegible]	[illegible]	[illegible]	[illegible]	[illegible]	[illegible]	[illegible]	[illegible]	[illegible]	[illegible]
[illegible]	[illegible]	[illegible]	[illegible]	[illegible]	[illegible]	[illegible]	[illegible]	[illegible]	[illegible]	[illegible]	[illegible]	[illegible]	[illegible]
[illegible]	[illegible]	[illegible]	[illegible]	[illegible]	[illegible]	[illegible]	[illegible]	[illegible]	[illegible]	[illegible]	[illegible]	[illegible]	[illegible]
[illegible]	[illegible]	[illegible]	[illegible]	[illegible]	[illegible]	[illegible]	[illegible]	[illegible]	[illegible]	[illegible]	[illegible]	[illegible]	[illegible]
[illegible]	[illegible]	[illegible]	[illegible]	[illegible]	[illegible]	[illegible]	[illegible]	[illegible]	[illegible]	[illegible]	[illegible]	[illegible]	[illegible]
[illegible]	[illegible]	[illegible]	[illegible]	[illegible]	[illegible]	[illegible]	[illegible]	[illegible]	[illegible]	[illegible]	[illegible]	[illegible]	[illegible]

From the foregoing it is evident that some of the younger animals included in the Statistics of Exports must of necessity escape enumeration in June of each year when the returns of live stock are collected for this Department. Viewing the number of animals exported in relation to those enumerated it is found that in cattle the number exported bear a relation of 16·4 per cent. to those enumerated in 1889, as compared with 18·0 per cent. in 1888; in sheep 16·2 per cent. as compared with 17·0 per cent. in 1888; and in pigs 34·3 per cent. as compared with 39·0 per cent. in 1888. Exports of Live Stock.

From the same Report it appears that the number of horses exported in 1889 amounted to 31,824, equal to 5·5 per cent. of those enumerated.

It also appears that during the same period there were imported into Ireland, 9,996 horses, 468 cattle, 39,791 sheep, and 181 pigs. Imports of Live Stock.

HONEY PRODUCED IN 1888.

In connexion with the Agricultural Statistics for 1889, Returns were obtained of the amount of Honey produced in the year 1888, and of the number of swarms at work. Tables compiled from the information contained in these Returns are given in the Produce Report presented to Parliament in January of this year. Honey produced in 1888.

SCUTCHING MILLS.

The number of Mills for scutching Flax in Ireland in 1889 was 1,062, being a decrease of 8 compared with 1888, and a decrease of 120 since the year 1880. 1,048 of these Mills in 1889 were in Ulster, 4 in Munster, 3 in Connaught, and 7 in Leinster. There were 447 Mills with from 1 to 4 stocks; 327 having 5 or 6; 253 with from 7 to 12; 33 having from 13 to 18; and 2 having above 18 stocks; 867 were worked by water power; 123 by steam; 71 by water and steam; and 1 by horse power. Scutching Mills 1889.

The following is the number of Scutching Mills, in each year, from 1880 to 1889 inclusive, by Provinces:— Scutching Mills, 1880 to 1889.

Provinces.	1880.	1881.	1882.	1883.	1884.	1885.	1886.	1887.	1888.	1889.
Leinster, . .	6	9	7	8	9	7	7	7	6	7
Munster, . .	18	15	19	15	12	9	8	6	4	4
Ulster, .	1,140	1,135	1,114	1,098	1,088	1,067	1,033	1,063	1,054	1,048
Connaught, . .	16	18	19	10	8	9	5	3	5	3
IRELAND, .	1,182	1,178	1,159	1,139	1,113	1,092	1,053	1,079	1,070	1,062

Scutching Mills, 1869. Number of Scutching Mills in 1869, by Counties and Provinces, classified according to the number of Stocks in each Mill, and the Power used in working them:—

Provinces and Counties in which there are Scutching Mills.	Classification of Mills.					Total No. of Mills.	Power Employed.				
	[illegible]	[illegible]	[illegible]	[illegible]	[illegible]		Water.	Steam.	Water and Steam.	Horse.	Wind.
Leinster:											
Longford,	.	.	.	1	.	1	1	.	.	.	.
Louth & Drogheda, Co. of Town,	.	.	4	.	.	4	3	1	.	.	.
Meath,	.	1	1	.	.	2	2	.	.	.	.
Total,	.	1	5	1	.	7	6	2	.	.	.
Munster:											
Cork,	4	.	.	.	.	4	4	.	.	.	.
Total,	4	.	.	.	.	4	4	.	.	.	.
Ulster:											
Antrim,	43	41	33	2	.	128	119	6	3	.	.
Armagh,	13	34	39	8	8	93	70	17	6	1	.
Cavan,	6	20	14	1	.	43	35	7	1	.	.
Donegal,	120	25	13	.	.	157	148	3	6	.	.
Down,	31	48	80	17	.	176	109	45	22	.	.
Fermanagh,	8	9	6	2	.	25	23	2	1	.	.
Londonderry,	96	58	19	.	.	173	158	5	10	.	.
Monaghan,	24	24	23	2	.	73	46	19	8	.	.
Tyrone,	88	48	38	3	.	177	133	31	10	.	.
Total,	443	328	346	33	8	1,048	856	120	71	1	.
Connaught:											
Leitrim,	.	.	1	.	.	1	1	.	.	.	.
Mayo,	.	1	1	.	.	2	1	1	.	.	.
Total,	.	1	2	.	.	3	2	1	.	.	.
Total of Ireland,	447	337	353	35	8	1,063	867	123	71	1	.

SILOS AND ENSILAGE.

Following the course adopted in the two previous years, in view of the interest attaching to the question of Ensilage, I communicated with those Landed Proprietors and Landholders, throughout the country, having Silos or otherwise making Ensilage, requesting them to favour me with certain details regarding the methods followed and the results obtained in the year 1889. I received replies to 340 out of 444 circulars issued by me, and I beg to express my obligations to my correspondents for the valuable and interesting information afforded. It will be found set forth in the Appendix, pp. 70 to 123.

The following Table shows, by Counties and Provinces, the number of Silos or Stacks mentioned in the communications received from the persons who forwarded replies to the circular above referred to. A reference to the communications shows that the decline in the number for last year, as compared with that for 1888, is attributable to the fact that the weather of 1889 was much more favourable for hay-making:—

Counties.	Number in 1888.	Number in 1889.	Counties.	Number in 1888.	Number in 1889.
Antrim,	53	27	Mayo,	10	18
Armagh,	3	1	Meath,	71	63
Carlow,	11	6	Monaghan,	4	9
Cavan,	13	1	Queen's,	37	19
Clare,	14	8	Roscommon,	3	3
			Sligo,	2	3
Cork,	49	21			
Donegal,	8	13	Tipperary,	55	24
Down,	10	8	Tyrone,	7	15
Dublin,	12	10	Waterford,	8	10
Fermanagh,	9	8	Westmeath,	19	4
			Wexford,	8	10
Galway,	23	18	Wicklow,	16	10
Kerry,	6	9			
Kildare,	9	14			
Kilkenny,	14	13	PROVINCES.		
King's,	35	22			
			Leinster,	240	175
Leitrim,	13	8	Munster,	120	91
Limerick,	13	15	Ulster,	120	95
Londonderry,	29	21	Connaught,	53	43
Longford,	5	6			
Louth,	8	5	TOTAL OF IRELAND,	533	404

In conclusion I have to thank the occupiers and owners of land in general for their courtesy in supplying the information required for the various Agricultural Returns to the Enumerators. I have also to express my thanks to the District Inspectors of the Royal Irish Constabulary and the Sergeants of the Metropolitan Police, who have furnished the valuable notes on the local circumstances affecting agriculture in the various parts of the country, which will be found at pages 60 to 69; and to add, as I do, with much pleasure, that the Enumerators discharged their duty with their usual efficiency.

I have the honour to remain,

Your Excellency's faithful servant,

T. W. GRIMSHAW.

Registrar-General.

General Register Office,

Charlemont House, Dublin,

19th May, 1890.

TILLAGE; MEADOW AND CLOVER; &c.

TABLE 1.—Showing, by Counties and Provinces, the Number of Holdings, their Size in Statute Acres, and the Division of Land in the Year 1889.

TABLE 2.—Showing the Percentage per cent. under Crops (including Meadow and Clover), Grass, Fallow, Woods and Plantations, Bog and Marsh, Barren Mountain Land, and Water, Roads, and Fences, &c., in each County and Province in Ireland in 1889.

[illegible] Poor Law Unions, the Number of [illegible] in [illegible] the Year 1889.

Poor Law Unions, the Number of Holdings, their Size in Statute Acres, and the Division of Land in the Year 1880—continued.

Holdings and their Size in Statute Acres							Total Number of Holdings	Division of Land							Total
[illegible]	[illegible]	[illegible]	[illegible]	[illegible]	[illegible]	[illegible]		[illegible]	[illegible]	[illegible]	[illegible]	[illegible]	[illegible]	[illegible]	
[illegible]	[illegible]	[illegible]	[illegible]	[illegible]	[illegible]	[illegible]	[illegible]	[illegible]	[illegible]	[illegible]	[illegible]	[illegible]	[illegible]	[illegible]	[illegible]

TABLE I.—SHOWING, BY COUNTIES AND PROVINCES, THE EXTENT OF LAND

COUNTIES.	CORN, BEANS, AND PEAS.								GREEN CROPS.		
	Wheat.	Oats.	Barley.	Bere.	Rye.	Beans.	Peas.	Total.	Potatoes.	Turnips.	Mangel Wurzel and Beet Root.
	Acres.	Acres.	Acres.	Acres.	Acres.	Acres.	Acres.	Acres.	Acres.	Acres.	Acres.
Antrim,	[illegible]	[illegible]	[illegible]	[illegible]	[illegible]	[illegible]	[illegible]	[illegible]	[illegible]	[illegible]	[illegible]
Armagh,	[illegible]	[illegible]	[illegible]	[illegible]	[illegible]	[illegible]	[illegible]	[illegible]	[illegible]	[illegible]	[illegible]
Carlow,	[illegible]	[illegible]	[illegible]	[illegible]	-	.	.	[illegible]	[illegible]	[illegible]	[illegible]
Cavan,	[illegible]	[illegible]	[illegible]	[illegible]	[illegible]	[illegible]	[illegible]	[illegible]	[illegible]	[illegible]	[illegible]
Clare,	[illegible]	[illegible]	[illegible]	[illegible]	[illegible]	[illegible]	[illegible]	[illegible]	[illegible]	[illegible]	[illegible]
Cork,	[illegible]	[illegible]	[illegible]	[illegible]	[illegible]	[illegible]	[illegible]	[illegible]	[illegible]	[illegible]	[illegible]
Donegal,	[illegible]	[illegible]	[illegible]	[illegible]	[illegible]	[illegible]	[illegible]	[illegible]	[illegible]	[illegible]	[illegible]
Down,	[illegible]	[illegible]	[illegible]	[illegible]	[illegible]	[illegible]	[illegible]	[illegible]	[illegible]	[illegible]	[illegible]
Dublin,	[illegible]	[illegible]	[illegible]	[illegible]	[illegible]	[illegible]	[illegible]	[illegible]	[illegible]	[illegible]	[illegible]
Fermanagh,	[illegible]	[illegible]	[illegible]	[illegible]	[illegible]	[illegible]	[illegible]	[illegible]	[illegible]	[illegible]	[illegible]
Galway,	[illegible]	[illegible]	[illegible]	[illegible]	[illegible]	[illegible]	[illegible]	[illegible]	[illegible]	[illegible]	[illegible]
Kerry,	[illegible]	[illegible]	[illegible]	[illegible]	[illegible]	[illegible]	.	[illegible]	[illegible]	[illegible]	[illegible]
Kildare,	[illegible]	[illegible]	[illegible]	[illegible]	[illegible]	[illegible]	[illegible]	[illegible]	[illegible]	[illegible]	[illegible]
Kilkenny,	[illegible]	[illegible]	[illegible]	.	[illegible]	.	.	[illegible]	[illegible]	[illegible]	[illegible]
King's,	[illegible]	[illegible]	[illegible]	[illegible]	[illegible]	[illegible]	[illegible]	[illegible]	[illegible]	[illegible]	[illegible]
Leitrim,	[illegible]	[illegible]	[illegible]	[illegible]	[illegible]	[illegible]	.	[illegible]	[illegible]	[illegible]	[illegible]
Limerick,	[illegible]	[illegible]	[illegible]	[illegible]	[illegible]	[illegible]	[illegible]	[illegible]	[illegible]	[illegible]	[illegible]
Londonderry,	[illegible]	[illegible]	[illegible]	[illegible]	[illegible]	[illegible]	[illegible]	[illegible]	[illegible]	[illegible]	[illegible]
Longford,	[illegible]	[illegible]	[illegible]	[illegible]	[illegible]	.	.	[illegible]	[illegible]	[illegible]	[illegible]
Louth and Drogheda, County of Town,	[illegible]	[illegible]	[illegible]	[illegible]	[illegible]	[illegible]	[illegible]	[illegible]	[illegible]	[illegible]	[illegible]
Mayo,	[illegible]	[illegible]	[illegible]	[illegible]	[illegible]	[illegible]	[illegible]	[illegible]	[illegible]	[illegible]	[illegible]
Meath,	[illegible]	[illegible]	[illegible]	[illegible]	[illegible]	[illegible]	.	[illegible]	[illegible]	[illegible]	[illegible]
Monaghan,	[illegible]	[illegible]	[illegible]	[illegible]	[illegible]	[illegible]	[illegible]	[illegible]	[illegible]	[illegible]	[illegible]
Queen's,	[illegible]	[illegible]	[illegible]	[illegible]	[illegible]	[illegible]	[illegible]	[illegible]	[illegible]	[illegible]	[illegible]
Roscommon,	[illegible]	[illegible]	[illegible]	[illegible]	[illegible]	[illegible]	.	[illegible]	[illegible]	[illegible]	[illegible]
Sligo,	[illegible]	[illegible]	[illegible]	[illegible]	[illegible]	[illegible]	[illegible]	[illegible]	[illegible]	[illegible]	[illegible]
Tipperary,	[illegible]	[illegible]	[illegible]	[illegible]	[illegible]	[illegible]	[illegible]	[illegible]	[illegible]	[illegible]	[illegible]
Tyrone,	[illegible]	[illegible]	[illegible]	[illegible]	[illegible]	[illegible]	[illegible]	[illegible]	[illegible]	[illegible]	[illegible]
Waterford,	[illegible]	[illegible]	[illegible]	[illegible]	[illegible]	.	.	[illegible]	[illegible]	[illegible]	[illegible]
Westmeath,	[illegible]	[illegible]	[illegible]	[illegible]	[illegible]	[illegible]	[illegible]	[illegible]	[illegible]	[illegible]	[illegible]
Wexford,	[illegible]	[illegible]	[illegible]	[illegible]	[illegible]	[illegible]	[illegible]	[illegible]	[illegible]	[illegible]	[illegible]
Wicklow,	[illegible]	[illegible]	[illegible]	[illegible]	.	.	.	[illegible]	[illegible]	[illegible]	[illegible]
PROVINCES.											
Leinster,	[illegible]	[illegible]	[illegible]	[illegible]	[illegible]	[illegible]	[illegible]	[illegible]	[illegible]	[illegible]	[illegible]
Munster,	[illegible]	[illegible]	[illegible]	[illegible]	[illegible]	[illegible]	[illegible]	[illegible]	[illegible]	[illegible]	[illegible]
Ulster,	[illegible]	[illegible]	[illegible]	[illegible]	[illegible]	[illegible]	[illegible]	[illegible]	[illegible]	[illegible]	[illegible]
Connaught,	[illegible]	[illegible]	[illegible]	[illegible]	[illegible]	[illegible]	[illegible]	[illegible]	[illegible]	[illegible]	[illegible]
Total,	[illegible]	[illegible]	[illegible]	[illegible]	[illegible]	[illegible]	[illegible]	[illegible]	[illegible]	[illegible]	[illegible]

[illegible] Crops in the Year 1869, the Valuation in 1869, and the Population in 1861.

[illegible]	[illegible]	[illegible]	[illegible]	[illegible]	Totals.	Flax.	[illegible]	[illegible]	[illegible]	[illegible]	[illegible]	Counties.
[illegible]	[illegible]	[illegible]	[illegible]	[illegible]	[illegible]	[illegible]	[illegible]	[illegible]	[illegible]	[illegible]	[illegible]	Antrim.
[illegible]	[illegible]	[illegible]	[illegible]	[illegible]	[illegible]	[illegible]	[illegible]	[illegible]	[illegible]	[illegible]	[illegible]	Armagh.
[illegible]	[illegible]	[illegible]	[illegible]	[illegible]	[illegible]		[illegible]	[illegible]	[illegible]	[illegible]	[illegible]	Carlow.
[illegible]	[illegible]	[illegible]	[illegible]	[illegible]	[illegible]	[illegible]	[illegible]	[illegible]	[illegible]	[illegible]	[illegible]	Cavan.
[illegible]	[illegible]	[illegible]	[illegible]	[illegible]	[illegible]	[illegible]	[illegible]	[illegible]	[illegible]	[illegible]	[illegible]	Clare.
[illegible]	[illegible]	[illegible]	[illegible]	[illegible]	[illegible]	[illegible]	[illegible]	[illegible]	[illegible]	[illegible]	[illegible]	Cork.
[illegible]	[illegible]	[illegible]	[illegible]	[illegible]	[illegible]	[illegible]	[illegible]	[illegible]	[illegible]	[illegible]	[illegible]	Donegal.
[illegible]	[illegible]	[illegible]	[illegible]	[illegible]	[illegible]	[illegible]	[illegible]	[illegible]	[illegible]	[illegible]	[illegible]	Down.
[illegible]	[illegible]	[illegible]	[illegible]	[illegible]	[illegible]	[illegible]	[illegible]	[illegible]	[illegible]	[illegible]	[illegible]	Dublin.
[illegible]	[illegible]	[illegible]	[illegible]	[illegible]	[illegible]	[illegible]	[illegible]	[illegible]	[illegible]	[illegible]	[illegible]	Fermanagh.
[illegible]	[illegible]	[illegible]	[illegible]	[illegible]	[illegible]	[illegible]	[illegible]	[illegible]	[illegible]	[illegible]	[illegible]	Galway.
[illegible]	[illegible]	[illegible]	[illegible]	[illegible]	[illegible]	[illegible]	[illegible]	[illegible]	[illegible]	[illegible]	[illegible]	Kerry.
[illegible]	[illegible]	[illegible]	[illegible]	[illegible]	[illegible]	[illegible]	[illegible]	[illegible]	[illegible]	[illegible]	[illegible]	Kildare.
[illegible]	[illegible]	[illegible]	[illegible]	[illegible]	[illegible]	[illegible]	[illegible]	[illegible]	[illegible]	[illegible]	[illegible]	Kilkenny.
[illegible]	[illegible]	[illegible]	[illegible]	[illegible]	[illegible]	[illegible]	[illegible]	[illegible]	[illegible]	[illegible]	[illegible]	King's.
[illegible]	[illegible]	[illegible]	[illegible]	[illegible]	[illegible]	[illegible]	[illegible]	[illegible]	[illegible]	[illegible]	[illegible]	Leitrim.
[illegible]	[illegible]	[illegible]	[illegible]	[illegible]	[illegible]	[illegible]	[illegible]	[illegible]	[illegible]	[illegible]	[illegible]	Limerick.
[illegible]	[illegible]	[illegible]	[illegible]	[illegible]	[illegible]	[illegible]	[illegible]	[illegible]	[illegible]	[illegible]	[illegible]	Londonderry.
[illegible]	[illegible]	[illegible]	[illegible]	[illegible]	[illegible]	[illegible]	[illegible]	[illegible]	[illegible]	[illegible]	[illegible]	Longford.
[illegible]	[illegible]	[illegible]	[illegible]	[illegible]	[illegible]	[illegible]	[illegible]	[illegible]	[illegible]	[illegible]	[illegible]	Louth & [illegible], County of Town.
[illegible]	[illegible]	[illegible]	[illegible]	[illegible]	[illegible]	[illegible]	[illegible]	[illegible]	[illegible]	[illegible]	[illegible]	Mayo.
[illegible]	[illegible]	[illegible]	[illegible]	[illegible]	[illegible]	[illegible]	[illegible]	[illegible]	[illegible]	[illegible]	[illegible]	Meath.
[illegible]	[illegible]	[illegible]	[illegible]	[illegible]	[illegible]	[illegible]	[illegible]	[illegible]	[illegible]	[illegible]	[illegible]	Monaghan.
[illegible]	[illegible]	[illegible]	[illegible]	[illegible]	[illegible]		[illegible]	[illegible]	[illegible]	[illegible]	[illegible]	Queen's.
[illegible]	[illegible]	[illegible]	[illegible]	[illegible]	[illegible]	[illegible]	[illegible]	[illegible]	[illegible]	[illegible]	[illegible]	Roscommon.
[illegible]	[illegible]	[illegible]	[illegible]	[illegible]	[illegible]	[illegible]	[illegible]	[illegible]	[illegible]	[illegible]	[illegible]	Sligo.
[illegible]	[illegible]	[illegible]	[illegible]	[illegible]	[illegible]		[illegible]	[illegible]	[illegible]	[illegible]	[illegible]	Tipperary.
[illegible]	[illegible]	[illegible]	[illegible]	[illegible]	[illegible]	[illegible]	[illegible]	[illegible]	[illegible]	[illegible]	[illegible]	Tyrone.
[illegible]	[illegible]	[illegible]	[illegible]	[illegible]	[illegible]		[illegible]	[illegible]	[illegible]	[illegible]	[illegible]	Waterford.
[illegible]	[illegible]	[illegible]	[illegible]	[illegible]	[illegible]		[illegible]	[illegible]	[illegible]	[illegible]	[illegible]	Westmeath.
[illegible]	[illegible]	[illegible]	[illegible]	[illegible]	[illegible]	[illegible]	[illegible]	[illegible]	[illegible]	[illegible]	[illegible]	Wexford.
[illegible]	[illegible]	[illegible]	[illegible]	[illegible]	[illegible]		[illegible]	[illegible]	[illegible]	[illegible]	[illegible]	Wicklow.
												PROVINCES.
[illegible]	[illegible]	[illegible]	[illegible]	[illegible]	[illegible]	[illegible]	[illegible]	[illegible]	[illegible]	[illegible]	[illegible]	Leinster.
[illegible]	[illegible]	[illegible]	[illegible]	[illegible]	[illegible]	[illegible]	[illegible]	[illegible]	[illegible]	[illegible]	[illegible]	Munster.
[illegible]	[illegible]	[illegible]	[illegible]	[illegible]	[illegible]	[illegible]	[illegible]	[illegible]	[illegible]	[illegible]	[illegible]	Ulster.
[illegible]	[illegible]	[illegible]	[illegible]	[illegible]	[illegible]	[illegible]	[illegible]	[illegible]	[illegible]	[illegible]	[illegible]	Connaught.
[illegible]	[illegible]	[illegible]	[illegible]	[illegible]	[illegible]	[illegible]	[illegible]	[illegible]	[illegible]	[illegible]	[illegible]	Total.

* [illegible]

Table 6.—Showing, by Counties and Provinces the

PRODUCE.						
Corn, Beans, and Peas.						
Wheat.	Oats.	Barley.	Bere.	Rye.	Beans.	Peas.
Cwts. of 112 lbs.	Cwts. of 112 lbs.	Cwts. of 112 lbs.	Cwts. of 112 lbs.	Cwts. of 112 lbs.	Cwts. of 112 lbs.	Cwts. of 112 lbs.
87,001	1,374,862	13,809	1,257	388	106,834	611
23,564	614,900	444	153	9,330	9,464	391
[illegible]	[illegible]	[illegible]	11	.	.	.
6,421	430,460	1,087	147	2,150	86	84
[illegible]	[illegible]	11,270	94	11,500	504	94
204,338	1,241,878	400,378	330	8,612	620	96
6,503	1,199,345	34,454	600	18,875	3,663	1,500
191,748	1,101,373	4,815	220	220	3,866	800
85,213	982,394	44,154	366	847	132	1,063
12,719	200,377	420	118	6,641	1389	114
27,433	837,760	80,736	616	34,813	70	434
33,463	546,412	33,448	298	7,088	343	.
28,142	841,019	108,533	414	1,344	394	36
150,418	401,841	400,568	..	390	.	.
2,302	609,071	501,602	360	12,193	31	14
8,663	156,436	14	40	4,305	18	.
36,638	300,807	13,635	32	673	133	17
18,425	1,068,438	72,232	273	3,533	8,840	462
8,184	668,146	313	33	4,448	.	.
72,870	600,371	366,550	24	113	270	466
33,390	722,283	18,430	354	46,611	64	446
95,964	878,108	93,081	70	4,073	334	.
10,059	403,300	12,660	758	394	364	22
11,393	745,112	423,480	138	334	31	16
6,361	591,438	4,166	144	19,140	139	.
10,651	617,350	8,673	36	8,936	46	14
124,565	438,896	305,248	287	5,331	115	36
51,063	1,202,178	414	96	3,320	202	79
30,821	471,340	38,186	40	430	.	.
949	538,308	4,480	77	3,374	99	21
61,554	710,356	488,333	488	384	87,610	43
10,954	518,000	8,800	38	.	.	.

PRODUCE OF THE CROPS IN THE YEAR 18[illegible].

[illegible]	[illegible]	[illegible]	[illegible]	[illegible]	[illegible]	[illegible]	[illegible]	HAY. [illegible]	HAY. [illegible]	COUNTIES.
[illegible]	[illegible]	[illegible]	[illegible]	[illegible]	[illegible]	[illegible]	[illegible]	[illegible]	[illegible]	
[illegible]	[illegible]	[illegible]	[illegible]	[illegible]	[illegible]	[illegible]	[illegible]	[illegible]	[illegible]	ANTRIM.
[illegible]	[illegible]	[illegible]	[illegible]	[illegible]	[illegible]	[illegible]	[illegible]	[illegible]	[illegible]	ARMAGH.
[illegible]	[illegible]	[illegible]	[illegible]	[illegible]	[illegible]	[illegible]	.	[illegible]	[illegible]	CARLOW.
[illegible]	[illegible]	[illegible]	[illegible]	[illegible]	[illegible]	[illegible]	[illegible]	[illegible]	[illegible]	CAVAN.
[illegible]	[illegible]	[illegible]	[illegible]	[illegible]	[illegible]	[illegible]	[illegible]	[illegible]	[illegible]	CLARE.
[illegible]	[illegible]	[illegible]	[illegible]	[illegible]	[illegible]	[illegible]	[illegible]	[illegible]	[illegible]	CORK.
[illegible]	[illegible]	[illegible]	[illegible]	[illegible]	[illegible]	[illegible]	[illegible]	[illegible]	[illegible]	DONEGAL.
[illegible]	[illegible]	[illegible]	[illegible]	[illegible]	[illegible]	[illegible]	[illegible]	[illegible]	[illegible]	DOWN.
[illegible]	[illegible]	[illegible]	[illegible]	[illegible]	[illegible]	[illegible]	.	[illegible]	[illegible]	DUBLIN.
[illegible]	[illegible]	[illegible]	[illegible]	[illegible]	[illegible]	[illegible]	[illegible]	[illegible]	[illegible]	FERMANAGH.
[illegible]	[illegible]	[illegible]	[illegible]	[illegible]	[illegible]	[illegible]	[illegible]	[illegible]	[illegible]	GALWAY.
[illegible]	[illegible]	[illegible]	[illegible]	[illegible]	[illegible]	[illegible]	[illegible]	[illegible]	[illegible]	KERRY.
[illegible]	[illegible]	[illegible]	[illegible]	[illegible]	[illegible]	[illegible]	.	[illegible]	[illegible]	KILDARE.
[illegible]	[illegible]	[illegible]	[illegible]	[illegible]	[illegible]	[illegible]	[illegible]	[illegible]	[illegible]	KILKENNY.
[illegible]	[illegible]	[illegible]	[illegible]	[illegible]	[illegible]	[illegible]	[illegible]	[illegible]	[illegible]	KING'S.
[illegible]	[illegible]	[illegible]	[illegible]	[illegible]	[illegible]	[illegible]	[illegible]	[illegible]	[illegible]	LEITRIM.
[illegible]	[illegible]	[illegible]	[illegible]	[illegible]	[illegible]	[illegible]	[illegible]	[illegible]	[illegible]	LIMERICK.
[illegible]	[illegible]	[illegible]	[illegible]	[illegible]	[illegible]	[illegible]	[illegible]	[illegible]	[illegible]	LONDONDERRY.
[illegible]	[illegible]	[illegible]	[illegible]	[illegible]	[illegible]	[illegible]	[illegible]	[illegible]	[illegible]	LONGFORD.
[illegible]	[illegible]	[illegible]	[illegible]	[illegible]	[illegible]	[illegible]	[illegible]	[illegible]	[illegible]	LOUTH and DROGHEDA, County of Town.
[illegible]	[illegible]	[illegible]	[illegible]	[illegible]	[illegible]	[illegible]	[illegible]	[illegible]	[illegible]	MAYO.
[illegible]	[illegible]	[illegible]	[illegible]	[illegible]	[illegible]	[illegible]	[illegible]	[illegible]	[illegible]	MEATH.
[illegible]	[illegible]	[illegible]	[illegible]	[illegible]	[illegible]	[illegible]	[illegible]	[illegible]	[illegible]	MONAGHAN.
[illegible]	[illegible]	[illegible]	[illegible]	[illegible]	[illegible]	[illegible]	.	[illegible]	[illegible]	QUEEN'S.
[illegible]	[illegible]	[illegible]	[illegible]	[illegible]	[illegible]	[illegible]	[illegible]	[illegible]	[illegible]	ROSCOMMON.
[illegible]	[illegible]	[illegible]	[illegible]	[illegible]	[illegible]	[illegible]	[illegible]	[illegible]	[illegible]	SLIGO.
[illegible]	[illegible]	[illegible]	[illegible]	[illegible]	[illegible]	[illegible]	.	[illegible]	[illegible]	TIPPERARY.
[illegible]	[illegible]	[illegible]	[illegible]	[illegible]	[illegible]	[illegible]	[illegible]	[illegible]	[illegible]	TYRONE.
[illegible]	[illegible]	[illegible]	[illegible]	[illegible]	[illegible]	[illegible]	.	[illegible]	[illegible]	WATERFORD.
[illegible]	[illegible]	[illegible]	[illegible]	[illegible]	[illegible]	[illegible]	.	[illegible]	[illegible]	WESTMEATH.
[illegible]	[illegible]	[illegible]	[illegible]	[illegible]	[illegible]	[illegible]	[illegible]	[illegible]	[illegible]	WEXFORD.
[illegible]	[illegible]	[illegible]	[illegible]	[illegible]	[illegible]	[illegible]	.	[illegible]	[illegible]	WICKLOW.
										PROVINCES.
[illegible]	[illegible]	[illegible]	[illegible]	[illegible]	[illegible]	[illegible]	[illegible]	[illegible]	[illegible]	LEINSTER.
[illegible]	[illegible]	[illegible]	[illegible]	[illegible]	[illegible]	[illegible]	[illegible]	[illegible]	[illegible]	MUNSTER.
[illegible]	[illegible]	[illegible]	[illegible]	[illegible]	[illegible]	[illegible]	[illegible]	[illegible]	[illegible]	ULSTER.
[illegible]	[illegible]	[illegible]	[illegible]	[illegible]	[illegible]	[illegible]	[illegible]	[illegible]	[illegible]	CONNAUGHT.
[illegible]	[illegible]	[illegible]	[illegible]	[illegible]	[illegible]	[illegible]	[illegible]	[illegible]	[illegible]	TOTAL.

[illegible]	[illegible]	[illegible]	[illegible]
[illegible]	[illegible]	[illegible]	[illegible]
[illegible]	[illegible]	[illegible]	[illegible]
[illegible]	[illegible]	[illegible]	[illegible]
[illegible]	[illegible]	[illegible]	[illegible]
[illegible]	[illegible]	[illegible]	[illegible]
[illegible]	[illegible]	[illegible]	[illegible]
[illegible]	[illegible]	[illegible]	[illegible]
[illegible]	[illegible]	[illegible]	[illegible]
[illegible]	[illegible]	[illegible]	[illegible]
[illegible]	[illegible]	[illegible]	[illegible]
[illegible]	[illegible]	[illegible]	[illegible]
[illegible]	[illegible]	[illegible]	[illegible]
[illegible]	[illegible]	[illegible]	[illegible]
[illegible]	[illegible]	[illegible]	[illegible]
[illegible]	[illegible]	[illegible]	[illegible]
[illegible]	[illegible]	[illegible]	[illegible]

TABLE 7.—Showing, by Poor Law Unions, the Extent of Land

Corn, Beans, and Peas.								Green Crops.		
Wheat.	Oats.	Barley.	Bere.	Rye.	Beans.	Peas.	Total.	Potatoes.	Turnips.	Mangel Wurzel and Beet Root.
Acres.	Acres.	Acres.	Acres.	Acres.	Acres.	Acres.	Acres.	Acres.	Acres.	Acres.
[illegible]	[illegible]	[illegible]	[illegible]	[illegible]	[illegible]	[illegible]	[illegible]	[illegible]	[illegible]	[illegible]

AGRICULTUBAL STATISTICS FOR THE YEAR 1869.

[illegible] in the Year 1869, [illegible] in 1868, and the [illegible] of 1861—continued.

[illegible]	[illegible]	[illegible]	[illegible]	[illegible]	[illegible]	[illegible]	[illegible]	[illegible]	[illegible]	[illegible]	[illegible]	POOR LAW UNIONS.
[illegible]	[illegible]	[illegible]	[illegible]	[illegible]	[illegible]	[illegible]	[illegible]	[illegible]	[illegible]	[illegible]	[illegible]	[illegible]
[illegible]	[illegible]	[illegible]	[illegible]	[illegible]	[illegible]	[illegible]	[illegible]	[illegible]	[illegible]	[illegible]	[illegible]	Total.

TABLE 5.—SHOWING, BY POOR LAW UNIONS, THE

POOR LAW UNIONS.	Wheat.	Oats.	Barley.	Bere.	Rye.	Beans.	Peas.
	[illegible]	[illegible]	[illegible]	[illegible]	[illegible]	[illegible]	[illegible]

[illegible] of the Crops in the Year 18[illegible]—continued.

[illegible]	[illegible]	[illegible]	[illegible]	[illegible]	[illegible]	[illegible]	[illegible]	[illegible]	[illegible]	POOR LAW UNIONS.
[illegible]	[illegible]	[illegible]	[illegible]	[illegible]	[illegible]	[illegible]	[illegible]	[illegible]	[illegible]	
[illegible]	[illegible]	[illegible]	[illegible]	[illegible]	[illegible]	[illegible]	[illegible]	[illegible]	[illegible]	[illegible]
[illegible]	[illegible]	[illegible]	[illegible]	[illegible]	[illegible]	[illegible]	[illegible]	[illegible]	[illegible]	Totals.

TABLE 8.—SHOWING THE NUMBER OF HOLDINGS EXCEEDING ONE ACRE, AND EXTENT OF LAND UNDER CROPS IN EACH YEAR FROM 1850 TO 1859, BY COUNTIES AND PROVINCES.

Counties	Years	Number of Holdings	Cereal Crops	Green Crops	Other Crops
Antrim. Area, [illegible] Acres.	[illegible]	[illegible]	[illegible]	[illegible]	[illegible]
Armagh. Area, [illegible] Acres.	[illegible]	[illegible]	[illegible]	[illegible]	[illegible]
Carlow. Area, [illegible] Acres.	[illegible]	[illegible]	[illegible]	[illegible]	[illegible]
Cavan. Area, [illegible] Acres.	[illegible]	[illegible]	[illegible]	[illegible]	[illegible]
Clare. Area, [illegible] Acres.	[illegible]	[illegible]	[illegible]	[illegible]	[illegible]
Cork. Area, [illegible] Acres.	[illegible]	[illegible]	[illegible]	[illegible]	[illegible]
Donegal. Area, [illegible] Acres.	[illegible]	[illegible]	[illegible]	[illegible]	[illegible]
Down. Area, [illegible] Acres.	[illegible]	[illegible]	[illegible]	[illegible]	[illegible]

[illegible]

TABLE [illegible].—Showing the Number of Holdings exceeding one Acre, and Extent of Land under Crops in each Year from 1880 to 1889, in Counties and Provinces—continued.

COUNTIES.	Years.	No. of Holdings exceeding 1 acre.	Extent under Crops in Statute Acres in each Year from 1880 to 1889 — Corn, Beans, and Peas							Green Crops							[illegible]
[illegible]	[illegible]	[illegible]	[illegible]	[illegible]	[illegible]	[illegible]	[illegible]	[illegible]	[illegible]	[illegible]	[illegible]	[illegible]	[illegible]	[illegible]	[illegible]	[illegible]	[illegible]

TABLE 6.—Showing the Number of Holdings exceeding one Acre, and Extent of Land under Crops in each Year from 1880 to 1889, by Counties and Provinces—continued.

PROVINCES.

Provinces.	Years.	No. of Holdings	Extent under Crops in Statute Acres in each Year from 1880 to 1889. — Corn, Beans, and Peas							Green Crops							Total under Crops.
Leinster. Area, [illegible] Acres.	[illegible]	[illegible]	[illegible]	[illegible]	[illegible]	[illegible]	[illegible]	[illegible]	[illegible]	[illegible]	[illegible]	[illegible]	[illegible]	[illegible]	[illegible]	[illegible]	[illegible]
Munster. Area, [illegible] Acres.	[illegible]	[illegible]	[illegible]	[illegible]	[illegible]	[illegible]	[illegible]	[illegible]	[illegible]	[illegible]	[illegible]	[illegible]	[illegible]	[illegible]	[illegible]	[illegible]	[illegible]
Ulster. Area, [illegible] Acres.	[illegible]	[illegible]	[illegible]	[illegible]	[illegible]	[illegible]	[illegible]	[illegible]	[illegible]	[illegible]	[illegible]	[illegible]	[illegible]	[illegible]	[illegible]	[illegible]	[illegible]
Connaught. Area, [illegible] Acres.	[illegible]	[illegible]	[illegible]	[illegible]	[illegible]	[illegible]	[illegible]	[illegible]	[illegible]	[illegible]	[illegible]	[illegible]	[illegible]	[illegible]	[illegible]	[illegible]	[illegible]

TOTAL OF IRELAND.

—	Years.	No. of Holdings	Extent under Crops in Statute Acres in each Year from 1880 to 1889. — Corn, Beans, and Peas							Green Crops							Total under Crops.
Total of Ireland. Area, [illegible] Acres.	[illegible]	[illegible]	[illegible]	[illegible]	[illegible]	[illegible]	[illegible]	[illegible]	[illegible]	[illegible]	[illegible]	[illegible]	[illegible]	[illegible]	[illegible]	[illegible]	[illegible]

TABLE 10.—SHOWING THE AVERAGE RATES OF PRODUCE OF CROPS TO THE STATUTE ACRE, IN EACH YEAR, FROM [illegible] to [illegible].

Counties.	Years.	Wheat.	Oats.	Barley.	Bere.	Rye.	Beans.	Peas.	Potatoes.	Turnips.	Mangel Wurzel and Beet-root.	Cabbage.	Flax.	Hay.
ANTRIM,	[illegible]	[illegible]	[illegible]	[illegible]	[illegible]	[illegible]	[illegible]	[illegible]	[illegible]	[illegible]	[illegible]	[illegible]	[illegible]	[illegible]
ARMAGH,	[illegible]	[illegible]	[illegible]	[illegible]	[illegible]	[illegible]	[illegible]	[illegible]	[illegible]	[illegible]	[illegible]	[illegible]	[illegible]	[illegible]
CARLOW,	[illegible]	[illegible]	[illegible]	[illegible]	[illegible]	[illegible]	[illegible]	[illegible]	[illegible]	[illegible]	[illegible]	[illegible]	[illegible]	[illegible]
CAVAN,	[illegible]	[illegible]	[illegible]	[illegible]	[illegible]	[illegible]	[illegible]	[illegible]	[illegible]	[illegible]	[illegible]	[illegible]	[illegible]	[illegible]
CLARE,	[illegible]	[illegible]	[illegible]	[illegible]	[illegible]	[illegible]	[illegible]	[illegible]	[illegible]	[illegible]	[illegible]	[illegible]	[illegible]	[illegible]
CORK,	[illegible]	[illegible]	[illegible]	[illegible]	[illegible]	[illegible]	[illegible]	[illegible]	[illegible]	[illegible]	[illegible]	[illegible]	[illegible]	[illegible]
DONEGAL,	[illegible]	[illegible]	[illegible]	[illegible]	[illegible]	[illegible]	[illegible]	[illegible]	[illegible]	[illegible]	[illegible]	[illegible]	[illegible]	[illegible]
DOWN,	[illegible]	[illegible]	[illegible]	[illegible]	[illegible]	[illegible]	[illegible]	[illegible]	[illegible]	[illegible]	[illegible]	[illegible]	[illegible]	[illegible]

F

TABLE 10.—ESTIMATED AVERAGE RATE OF PRODUCE [illegible]—continued.

Counties.	Years.	Wheat.	Oats.	Barley.	Bere.	Rye.	Beans.	Peas.	Potatoes.	Turnips.	Mangel Wurzel [illegible]	Cabbage.	Flax.	Hay.
		[illegible]	[illegible]	[illegible]	[illegible]	[illegible]	[illegible]	[illegible]	Tons.	Tons.	Tons.	Tons.	[illegible]	Tons.
Limerick,	[illegible]	[illegible]	[illegible]	[illegible]	[illegible]	[illegible]	[illegible]	[illegible]	[illegible]	[illegible]	[illegible]	[illegible]	[illegible]	[illegible]
Londonderry,	[illegible]	[illegible]	[illegible]	[illegible]	[illegible]	[illegible]	[illegible]	[illegible]	[illegible]	[illegible]	[illegible]	[illegible]	[illegible]	[illegible]
Longford,	[illegible]	[illegible]	[illegible]	[illegible]	[illegible]	[illegible]	[illegible]	[illegible]	[illegible]	[illegible]	[illegible]	[illegible]	[illegible]	[illegible]
Louth, [illegible]	[illegible]	[illegible]	[illegible]	[illegible]	[illegible]	[illegible]	[illegible]	[illegible]	[illegible]	[illegible]	[illegible]	[illegible]	[illegible]	[illegible]
Mayo,	[illegible]	[illegible]	[illegible]	[illegible]	[illegible]	[illegible]	[illegible]	[illegible]	[illegible]	[illegible]	[illegible]	[illegible]	[illegible]	[illegible]
Meath,	[illegible]	[illegible]	[illegible]	[illegible]	[illegible]	[illegible]	[illegible]	[illegible]	[illegible]	[illegible]	[illegible]	[illegible]	[illegible]	[illegible]
Monaghan,	[illegible]	[illegible]	[illegible]	[illegible]	[illegible]	[illegible]	[illegible]	[illegible]	[illegible]	[illegible]	[illegible]	[illegible]	[illegible]	[illegible]
Queen's,	[illegible]	[illegible]	[illegible]	[illegible]	[illegible]	[illegible]	[illegible]	[illegible]	[illegible]	[illegible]	[illegible]	[illegible]	[illegible]	[illegible]

[illegible]

TABLE 10.—SHOWING THE AVERAGE RATES OF PRODUCE TO THE STATUTE ACRE—*continued.*

AVERAGE OF PROVINCES.

Provinces.	Years.	Wheat.	Oats.	Barley.	Bere.	Rye.	Beans.	Peas.	Potatoes.	Turnips.	Mangel Wurzel and Beet-root.	Cabbage.	Flax.	Hay.
Leinster,	[illegible]	[illegible]	[illegible]	[illegible]	[illegible]	[illegible]	[illegible]	[illegible]	[illegible]	[illegible]	[illegible]	[illegible]	[illegible]	[illegible]
Munster,	[illegible]	[illegible]	[illegible]	[illegible]	[illegible]	[illegible]	[illegible]	[illegible]	[illegible]	[illegible]	[illegible]	[illegible]	[illegible]	[illegible]
Ulster,	[illegible]	[illegible]	[illegible]	[illegible]	[illegible]	[illegible]	[illegible]	[illegible]	[illegible]	[illegible]	[illegible]	[illegible]	[illegible]	[illegible]
Connaught,	[illegible]	[illegible]	[illegible]	[illegible]	[illegible]	[illegible]	[illegible]	[illegible]	[illegible]	[illegible]	[illegible]	[illegible]	[illegible]	[illegible]

AVERAGE OF IRELAND.

	Years.	Wheat.	Oats.	Barley.	Bere.	Rye.	Beans.	Peas.	Potatoes.	Turnips.	Mangel Wurzel and Beet-root.	Cabbage.	Flax.	Hay.
Ireland,	[illegible]	[illegible]	[illegible]	[illegible]	[illegible]	[illegible]	[illegible]	[illegible]	[illegible]	[illegible]	[illegible]	[illegible]	[illegible]	[illegible]

[illegible]

TABLE 18.—Showing the Quantity of Live Stock in each Year from 1860 to 1869, in Counties and Provinces.

Counties.	Years.	No. of Horses.			Mules and Asses.		No. of Cattle.			No. of Sheep.		No. of Pigs.		No. of Goats.	No. of Poultry.
		[illegible]	[illegible]	[illegible]	[illegible]	[illegible]	[illegible]	[illegible]	[illegible]	[illegible]	[illegible]	[illegible]	[illegible]		
Antrim: Area, [illegible] Acres.	[illegible]	[illegible]	[illegible]	[illegible]	[illegible]	[illegible]	[illegible]	[illegible]	[illegible]	[illegible]	[illegible]	[illegible]	[illegible]	[illegible]	[illegible]
Armagh: Area, [illegible] Acres.	[illegible]	[illegible]	[illegible]	[illegible]	[illegible]	[illegible]	[illegible]	[illegible]	[illegible]	[illegible]	[illegible]	[illegible]	[illegible]	[illegible]	[illegible]
Carlow: Area, [illegible] Acres.	[illegible]	[illegible]	[illegible]	[illegible]	[illegible]	[illegible]	[illegible]	[illegible]	[illegible]	[illegible]	[illegible]	[illegible]	[illegible]	[illegible]	[illegible]
Cavan: Area, [illegible] Acres.	[illegible]	[illegible]	[illegible]	[illegible]	[illegible]	[illegible]	[illegible]	[illegible]	[illegible]	[illegible]	[illegible]	[illegible]	[illegible]	[illegible]	[illegible]
Clare: Area, [illegible] Acres.	[illegible]	[illegible]	[illegible]	[illegible]	[illegible]	[illegible]	[illegible]	[illegible]	[illegible]	[illegible]	[illegible]	[illegible]	[illegible]	[illegible]	[illegible]
Cork: Area, [illegible] Acres.	[illegible]	[illegible]	[illegible]	[illegible]	[illegible]	[illegible]	[illegible]	[illegible]	[illegible]	[illegible]	[illegible]	[illegible]	[illegible]	[illegible]	[illegible]
Donegal: Area, [illegible] Acres.	[illegible]	[illegible]	[illegible]	[illegible]	[illegible]	[illegible]	[illegible]	[illegible]	[illegible]	[illegible]	[illegible]	[illegible]	[illegible]	[illegible]	[illegible]
Down: Area, [illegible] Acres.	[illegible]	[illegible]	[illegible]	[illegible]	[illegible]	[illegible]	[illegible]	[illegible]	[illegible]	[illegible]	[illegible]	[illegible]	[illegible]	[illegible]	[illegible]

TABLE 13.—SHOWING THE QUANTITY OF LIVE STOCK IN EACH YEAR FROM 1860 TO 1869, BY COUNTIES AND PROVINCES—continued.

COUNTIES.	Year.	No. of Horses			Mules and Asses		No. of Cattle			No. of Sheep		No. of Pigs		No. of Goats.	No. of Poultry.
		[illegible]	[illegible]	Under two years	No. of Mules.	No. of Asses.	[illegible]	[illegible]	Under two years	[illegible]	Under one year	[illegible]	Under one year		
[illegible] Area, [illegible] Acres.	[illegible]	[illegible]	[illegible]	[illegible]	[illegible]	[illegible]	[illegible]	[illegible]	[illegible]	[illegible]	[illegible]	[illegible]	[illegible]	[illegible]	[illegible]
[illegible] Area, [illegible] Acres.	[illegible]	[illegible]	[illegible]	[illegible]	[illegible]	[illegible]	[illegible]	[illegible]	[illegible]	[illegible]	[illegible]	[illegible]	[illegible]	[illegible]	[illegible]
[illegible] Area, [illegible] Acres.	[illegible]	[illegible]	[illegible]	[illegible]	[illegible]	[illegible]	[illegible]	[illegible]	[illegible]	[illegible]	[illegible]	[illegible]	[illegible]	[illegible]	[illegible]
[illegible] Area, [illegible] Acres.	[illegible]	[illegible]	[illegible]	[illegible]	[illegible]	[illegible]	[illegible]	[illegible]	[illegible]	[illegible]	[illegible]	[illegible]	[illegible]	[illegible]	[illegible]
[illegible] Area, [illegible] Acres.	[illegible]	[illegible]	[illegible]	[illegible]	[illegible]	[illegible]	[illegible]	[illegible]	[illegible]	[illegible]	[illegible]	[illegible]	[illegible]	[illegible]	[illegible]
[illegible] Area, [illegible] Acres.	[illegible]	[illegible]	[illegible]	[illegible]	[illegible]	[illegible]	[illegible]	[illegible]	[illegible]	[illegible]	[illegible]	[illegible]	[illegible]	[illegible]	[illegible]
[illegible] Area, [illegible] Acres.	[illegible]	[illegible]	[illegible]	[illegible]	[illegible]	[illegible]	[illegible]	[illegible]	[illegible]	[illegible]	[illegible]	[illegible]	[illegible]	[illegible]	[illegible]
[illegible] Area, [illegible] Acres.	[illegible]	[illegible]	[illegible]	[illegible]	[illegible]	[illegible]	[illegible]	[illegible]	[illegible]	[illegible]	[illegible]	[illegible]	[illegible]	[illegible]	[illegible]

Table 10.—Showing the Quantity of Live Stock in each Year from [illegible] to [illegible], by Counties and Provinces—continued.

Counties.	Years.	No. of Horses.			No. of Mules and Asses.		No. of Cattle.			No. of Sheep.		No. of Pigs.		[illegible]	[illegible]
		[illegible]	[illegible]	[illegible]	[illegible]	[illegible]	[illegible]	[illegible]	[illegible]	[illegible]	[illegible]	[illegible]	[illegible]		
[illegible]	[illegible]	[illegible]	[illegible]	[illegible]	[illegible]	[illegible]	[illegible]	[illegible]	[illegible]	[illegible]	[illegible]	[illegible]	[illegible]	[illegible]	[illegible]
[illegible]	[illegible]	[illegible]	[illegible]	[illegible]	[illegible]	[illegible]	[illegible]	[illegible]	[illegible]	[illegible]	[illegible]	[illegible]	[illegible]	[illegible]	[illegible]
[illegible]	[illegible]	[illegible]	[illegible]	[illegible]	[illegible]	[illegible]	[illegible]	[illegible]	[illegible]	[illegible]	[illegible]	[illegible]	[illegible]	[illegible]	[illegible]
[illegible]	[illegible]	[illegible]	[illegible]	[illegible]	[illegible]	[illegible]	[illegible]	[illegible]	[illegible]	[illegible]	[illegible]	[illegible]	[illegible]	[illegible]	[illegible]
[illegible]	[illegible]	[illegible]	[illegible]	[illegible]	[illegible]	[illegible]	[illegible]	[illegible]	[illegible]	[illegible]	[illegible]	[illegible]	[illegible]	[illegible]	[illegible]
[illegible]	[illegible]	[illegible]	[illegible]	[illegible]	[illegible]	[illegible]	[illegible]	[illegible]	[illegible]	[illegible]	[illegible]	[illegible]	[illegible]	[illegible]	[illegible]
[illegible]	[illegible]	[illegible]	[illegible]	[illegible]	[illegible]	[illegible]	[illegible]	[illegible]	[illegible]	[illegible]	[illegible]	[illegible]	[illegible]	[illegible]	[illegible]
[illegible]	[illegible]	[illegible]	[illegible]	[illegible]	[illegible]	[illegible]	[illegible]	[illegible]	[illegible]	[illegible]	[illegible]	[illegible]	[illegible]	[illegible]	[illegible]

TABLE 13.—SHOWING THE QUANTITY OF LIVE STOCK IN EACH YEAR FROM [illegible] TO [illegible], BY COUNTIES AND PROVINCES—*continued*.

COUNTIES.	Years.	No. of Horses.			Mules and Asses.		No. of Cattle.			No. of Sheep.		No. of Pigs.		[illegible]	[illegible]
		[illegible]	[illegible]	[illegible]	[illegible]	[illegible]	[illegible]	[illegible]	[illegible]	[illegible]	[illegible]	[illegible]	[illegible]		
[illegible]	[illegible]	[illegible]	[illegible]	[illegible]	[illegible]	[illegible]	[illegible]	[illegible]	[illegible]	[illegible]	[illegible]	[illegible]	[illegible]	[illegible]	[illegible]

TABLE 12.—SHOWING THE QUANTITY OF LIVE STOCK IN EACH YEAR FROM 1850 TO 1859, BY COUNTIES AND PROVINCES—*continued.*

PROVINCES.

PROVINCES.	Years.	No. of Horses: Two years old and upwards	No. of Horses: One year old and under two	No. of Horses: Under one year	Mules and Asses: No. of Mules	Mules and Asses: No. of Asses	No. of Cattle: Two years old and upwards	No. of Cattle: One year old and under two	No. of Cattle: Under one year	No. of Sheep: One year old and upwards	No. of Sheep: Under one year	No. of Pigs: One year old and upwards	No. of Pigs: Under one year	No. of Goats.	No. of Poultry.
LEINSTER: Area, [illegible] Acres.	[illegible]	[illegible]	[illegible]	[illegible]	[illegible]	[illegible]	[illegible]	[illegible]	[illegible]	[illegible]	[illegible]	[illegible]	[illegible]	[illegible]	[illegible]
MUNSTER: Area, [illegible] Acres.	[illegible]	[illegible]	[illegible]	[illegible]	[illegible]	[illegible]	[illegible]	[illegible]	[illegible]	[illegible]	[illegible]	[illegible]	[illegible]	[illegible]	[illegible]
ULSTER: Area, [illegible] Acres.	[illegible]	[illegible]	[illegible]	[illegible]	[illegible]	[illegible]	[illegible]	[illegible]	[illegible]	[illegible]	[illegible]	[illegible]	[illegible]	[illegible]	[illegible]
CONNAUGHT: Area, [illegible] Acres.	[illegible]	[illegible]	[illegible]	[illegible]	[illegible]	[illegible]	[illegible]	[illegible]	[illegible]	[illegible]	[illegible]	[illegible]	[illegible]	[illegible]	[illegible]

TOTAL OF IRELAND.

IRELAND.	Years.	No. of Horses: Two years old and upwards	No. of Horses: One year old and under two	No. of Horses: Under one year	Mules and Asses: No. of Mules	Mules and Asses: No. of Asses	No. of Cattle: Two years old and upwards	No. of Cattle: One year old and under two	No. of Cattle: Under one year	No. of Sheep: One year old and upwards	No. of Sheep: Under one year	No. of Pigs: One year old and upwards	No. of Pigs: Under one year	No. of Goats.	No. of Poultry.
Total of Ireland: Area, [illegible] Acres.	[illegible]	[illegible]	[illegible]	[illegible]	[illegible]	[illegible]	[illegible]	[illegible]	[illegible]	[illegible]	[illegible]	[illegible]	[illegible]	[illegible]	[illegible]

TABLE 14.—SHOWING, by COUNTIES and PROVINCES, the Total Area under POTATOES in 1880, and the Extent in Statute Acres under each description of that crop.

COUNTIES.	[illegible]	[illegible]	[illegible]	[illegible]	[illegible]	[illegible]	[illegible]	[illegible]	[illegible]	[illegible]	[illegible]	[illegible]	[illegible]	[illegible]	[illegible]	[illegible]	All others.
Antrim,	[illegible]	[illegible]	[illegible]	[illegible]	[illegible]	[illegible]	[illegible]	[illegible]	[illegible]	[illegible]	[illegible]	[illegible]	[illegible]	[illegible]	[illegible]	[illegible]	[illegible]
Armagh,	[illegible]	[illegible]	[illegible]	[illegible]	[illegible]	[illegible]	[illegible]	[illegible]	[illegible]	[illegible]	[illegible]	[illegible]	[illegible]	[illegible]	[illegible]	[illegible]	[illegible]
Carlow,	[illegible]	[illegible]	[illegible]	[illegible]	[illegible]	[illegible]	[illegible]	[illegible]	[illegible]	[illegible]	[illegible]	[illegible]	[illegible]	[illegible]	[illegible]	[illegible]	[illegible]
Cavan,	[illegible]	[illegible]	[illegible]	[illegible]	[illegible]	[illegible]	[illegible]	[illegible]	[illegible]	[illegible]	[illegible]	[illegible]	[illegible]	[illegible]	[illegible]	[illegible]	[illegible]
Clare,	[illegible]	[illegible]	[illegible]	[illegible]	[illegible]	[illegible]	[illegible]	[illegible]	[illegible]	[illegible]	[illegible]	[illegible]	[illegible]	[illegible]	[illegible]	[illegible]	[illegible]
Cork,	[illegible]	[illegible]	[illegible]	[illegible]	[illegible]	[illegible]	[illegible]	[illegible]	[illegible]	[illegible]	[illegible]	[illegible]	[illegible]	[illegible]	[illegible]	[illegible]	[illegible]
Donegal,	[illegible]	[illegible]	[illegible]	[illegible]	[illegible]	[illegible]	[illegible]	[illegible]	[illegible]	[illegible]	[illegible]	[illegible]	[illegible]	[illegible]	[illegible]	[illegible]	[illegible]
Down,	[illegible]	[illegible]	[illegible]	[illegible]	[illegible]	[illegible]	[illegible]	[illegible]	[illegible]	[illegible]	[illegible]	[illegible]	[illegible]	[illegible]	[illegible]	[illegible]	[illegible]
Dublin,	[illegible]	[illegible]	[illegible]	[illegible]	[illegible]	[illegible]	[illegible]	[illegible]	[illegible]	[illegible]	[illegible]	[illegible]	[illegible]	[illegible]	[illegible]	[illegible]	[illegible]
Fermanagh,	[illegible]	[illegible]	[illegible]	[illegible]	[illegible]	[illegible]	[illegible]	[illegible]	[illegible]	[illegible]	[illegible]	[illegible]	[illegible]	[illegible]	[illegible]	[illegible]	[illegible]
Galway,	[illegible]	[illegible]	[illegible]	[illegible]	[illegible]	[illegible]	[illegible]	[illegible]	[illegible]	[illegible]	[illegible]	[illegible]	[illegible]	[illegible]	[illegible]	[illegible]	[illegible]
Kerry,	[illegible]	[illegible]	[illegible]	[illegible]	[illegible]	[illegible]	[illegible]	[illegible]	[illegible]	[illegible]	[illegible]	[illegible]	[illegible]	[illegible]	[illegible]	[illegible]	[illegible]
Kildare,	[illegible]	[illegible]	[illegible]	[illegible]	[illegible]	[illegible]	[illegible]	[illegible]	[illegible]	[illegible]	[illegible]	[illegible]	[illegible]	[illegible]	[illegible]	[illegible]	[illegible]
Kilkenny,	[illegible]	[illegible]	[illegible]	[illegible]	[illegible]	[illegible]	[illegible]	[illegible]	[illegible]	[illegible]	[illegible]	[illegible]	[illegible]	[illegible]	[illegible]	[illegible]	[illegible]
King's,	[illegible]	[illegible]	[illegible]	[illegible]	[illegible]	[illegible]	[illegible]	[illegible]	[illegible]	[illegible]	[illegible]	[illegible]	[illegible]	[illegible]	[illegible]	[illegible]	[illegible]
Leitrim,	[illegible]	[illegible]	[illegible]	[illegible]	[illegible]	[illegible]	[illegible]	[illegible]	[illegible]	[illegible]	[illegible]	[illegible]	[illegible]	[illegible]	[illegible]	[illegible]	[illegible]
Limerick,	[illegible]	[illegible]	[illegible]	[illegible]	[illegible]	[illegible]	[illegible]	[illegible]	[illegible]	[illegible]	[illegible]	[illegible]	[illegible]	[illegible]	[illegible]	[illegible]	[illegible]
Londonderry,	[illegible]	[illegible]	[illegible]	[illegible]	[illegible]	[illegible]	[illegible]	[illegible]	[illegible]	[illegible]	[illegible]	[illegible]	[illegible]	[illegible]	[illegible]	[illegible]	[illegible]
Longford,	[illegible]	[illegible]	[illegible]	[illegible]	[illegible]	[illegible]	[illegible]	[illegible]	[illegible]	[illegible]	[illegible]	[illegible]	[illegible]	[illegible]	[illegible]	[illegible]	[illegible]
Louth and Drogheda, County of Town,	[illegible]	[illegible]	[illegible]	[illegible]	[illegible]	[illegible]	[illegible]	[illegible]	[illegible]	[illegible]	[illegible]	[illegible]	[illegible]	[illegible]	[illegible]	[illegible]	[illegible]
Mayo,	[illegible]	[illegible]	[illegible]	[illegible]	[illegible]	[illegible]	[illegible]	[illegible]	[illegible]	[illegible]	[illegible]	[illegible]	[illegible]	[illegible]	[illegible]	[illegible]	[illegible]
Meath,	[illegible]	[illegible]	[illegible]	[illegible]	[illegible]	[illegible]	[illegible]	[illegible]	[illegible]	[illegible]	[illegible]	[illegible]	[illegible]	[illegible]	[illegible]	[illegible]	[illegible]
Monaghan,	[illegible]	[illegible]	[illegible]	[illegible]	[illegible]	[illegible]	[illegible]	[illegible]	[illegible]	[illegible]	[illegible]	[illegible]	[illegible]	[illegible]	[illegible]	[illegible]	[illegible]
Queen's,	[illegible]	[illegible]	[illegible]	[illegible]	[illegible]	[illegible]	[illegible]	[illegible]	[illegible]	[illegible]	[illegible]	[illegible]	[illegible]	[illegible]	[illegible]	[illegible]	[illegible]
Roscommon,	[illegible]	[illegible]	[illegible]	[illegible]	[illegible]	[illegible]	[illegible]	[illegible]	[illegible]	[illegible]	[illegible]	[illegible]	[illegible]	[illegible]	[illegible]	[illegible]	[illegible]
Sligo,	[illegible]	[illegible]	[illegible]	[illegible]	[illegible]	[illegible]	[illegible]	[illegible]	[illegible]	[illegible]	[illegible]	[illegible]	[illegible]	[illegible]	[illegible]	[illegible]	[illegible]
Tipperary,	[illegible]	[illegible]	[illegible]	[illegible]	[illegible]	[illegible]	[illegible]	[illegible]	[illegible]	[illegible]	[illegible]	[illegible]	[illegible]	[illegible]	[illegible]	[illegible]	[illegible]
Tyrone,	[illegible]	[illegible]	[illegible]	[illegible]	[illegible]	[illegible]	[illegible]	[illegible]	[illegible]	[illegible]	[illegible]	[illegible]	[illegible]	[illegible]	[illegible]	[illegible]	[illegible]
Waterford,	[illegible]	[illegible]	[illegible]	[illegible]	[illegible]	[illegible]	[illegible]	[illegible]	[illegible]	[illegible]	[illegible]	[illegible]	[illegible]	[illegible]	[illegible]	[illegible]	[illegible]
Westmeath,	[illegible]	[illegible]	[illegible]	[illegible]	[illegible]	[illegible]	[illegible]	[illegible]	[illegible]	[illegible]	[illegible]	[illegible]	[illegible]	[illegible]	[illegible]	[illegible]	[illegible]
Wexford,	[illegible]	[illegible]	[illegible]	[illegible]	[illegible]	[illegible]	[illegible]	[illegible]	[illegible]	[illegible]	[illegible]	[illegible]	[illegible]	[illegible]	[illegible]	[illegible]	[illegible]
Wicklow,	[illegible]	[illegible]	[illegible]	[illegible]	[illegible]	[illegible]	[illegible]	[illegible]	[illegible]	[illegible]	[illegible]	[illegible]	[illegible]	[illegible]	[illegible]	[illegible]	[illegible]
PROVINCES.																	
Leinster,	[illegible]	[illegible]	[illegible]	[illegible]	[illegible]	[illegible]	[illegible]	[illegible]	[illegible]	[illegible]	[illegible]	[illegible]	[illegible]	[illegible]	[illegible]	[illegible]	[illegible]
Munster,	[illegible]	[illegible]	[illegible]	[illegible]	[illegible]	[illegible]	[illegible]	[illegible]	[illegible]	[illegible]	[illegible]	[illegible]	[illegible]	[illegible]	[illegible]	[illegible]	[illegible]
Ulster,	[illegible]	[illegible]	[illegible]	[illegible]	[illegible]	[illegible]	[illegible]	[illegible]	[illegible]	[illegible]	[illegible]	[illegible]	[illegible]	[illegible]	[illegible]	[illegible]	[illegible]
Connaught,	[illegible]	[illegible]	[illegible]	[illegible]	[illegible]	[illegible]	[illegible]	[illegible]	[illegible]	[illegible]	[illegible]	[illegible]	[illegible]	[illegible]	[illegible]	[illegible]	[illegible]
Total of Ireland, [illegible]	[illegible]	[illegible]	[illegible]	[illegible]	[illegible]	[illegible]	[illegible]	[illegible]	[illegible]	[illegible]	[illegible]	[illegible]	[illegible]	[illegible]	[illegible]	[illegible]	[illegible]
Percentage in [illegible]	[illegible]	[illegible]	[illegible]	[illegible]	[illegible]	[illegible]	[illegible]	[illegible]	[illegible]								
Total of Ireland, [illegible]	[illegible]	[illegible]	[illegible]	[illegible]	[illegible]	[illegible]	[illegible]	[illegible]	[illegible]	[illegible]	[illegible]	[illegible]	[illegible]	[illegible]	[illegible]	[illegible]	[illegible]
Percentage in [illegible]	[illegible]	[illegible]	[illegible]	[illegible]	[illegible]	[illegible]	[illegible]	[illegible]	[illegible]								

TABLE 13.—SHOWING, by POOR LAW UNIONS, the Total extent in STATUTE ACRES under POTATOES in 1889, and the extent under each description of that Crop.

TABLE 18.—Showing, by Poor Law Unions, the Total extent in Statute Acres under Potatoes in 1899, and the extent under each description of that Crop—*continued*.

TABLE 16.—SHOWING, by COUNTIES, the average rate of Produce per statute acre of the principal descriptions of POTATOES planted in Ireland in 1862.

COUNTIES.	GENERAL NAMES OF THE DIFFERENT KINDS OF POTATOES PLANTED IN EACH COUNTY.															
	[illegible]	[illegible]	[illegible]	[illegible]	[illegible]	[illegible]	[illegible]	[illegible]	[illegible]	[illegible]	[illegible]	[illegible]	[illegible]	[illegible]	[illegible]	[illegible]
	Cwts.	Cwts.	Cwts.	Cwts.	Cwts.	Cwts.	Cwts.	Cwts.	Cwts.	Cwts.	Cwts.	Cwts.	Cwts.	Cwts.	Cwts.	Cwts.
Antrim, . . .	[illegible]	71	[illegible]	47	[illegible]	[illegible]	[illegible]	[illegible]	[illegible]	[illegible]	.	.	.	.	[illegible]	.
Armagh, . . .	[illegible]	[illegible]	[illegible]	[illegible]	[illegible]	[illegible]	[illegible]	[illegible]	.	[illegible]	.	.	.	.	.	.
Carlow, . . .	[illegible]	74	.	[illegible]	[illegible]	[illegible]	[illegible]	.	.	[illegible]	.	.	.	.	.	.
Cavan, . . .	[illegible]	[illegible]	[illegible]	[illegible]	73	[illegible]	[illegible]	.	.	[illegible]	.	.	.	.	.	.
Clare, . . .	47	41	[illegible]	[illegible]	.	[illegible]	[illegible]	.	.	.	.	[illegible]	.	.	.	.
Cork, . . .	[illegible]	79	72	[illegible]	.	[illegible]	[illegible]	.	.	.	.	.	.	.	.	.
Donegal, . . .	74	[illegible]	[illegible]	79	[illegible]	[illegible]	[illegible]	.	75	71	[illegible]	.	.	.	.	71
Down, . . .	[illegible]	[illegible]	[illegible]	[illegible]	[illegible]	[illegible]	[illegible]	[illegible]	.	[illegible]	.	.	.	.	.	.
Dublin, . . .	[illegible]	[illegible]	[illegible]	[illegible]	70	[illegible]	[illegible]	.	.	.	.	.	.	.	.	.
Fermanagh, . . .	[illegible]	[illegible]	[illegible]	[illegible]	[illegible]	[illegible]	[illegible]	.	.	.	.	.	.	.	.	.
Galway, . . .	74	[illegible]	[illegible]	[illegible]	[illegible]	[illegible]	[illegible]	.	.	.	.	.	.	[illegible]	.	.
Kerry, . . .	[illegible]	[illegible]	[illegible]	[illegible]	.	[illegible]	[illegible]	.	.	.	.	[illegible]	.	.	.	.
Kildare, . . .	[illegible]	[illegible]	[illegible]	[illegible]	[illegible]	[illegible]	[illegible]	.	.	[illegible]	.	.	.	.	.	.
Kilkenny, . . .	76	[illegible]	[illegible]	71	[illegible]	[illegible]	[illegible]	.	.	[illegible]	.	.	.	.	.	.
King's, . . .	[illegible]	[illegible]	[illegible]	[illegible]	.	[illegible]	[illegible]	.	.	[illegible]	.	.	.	.	.	.
Leitrim, . . .	[illegible]	[illegible]	[illegible]	[illegible]	.	[illegible]	.	.	.	.	.	.	.	.	.	.
Limerick, . . .	[illegible]	71	[illegible]	73	.	[illegible]	[illegible]	.	.	.	.	[illegible]	.	.	.	.
Londonderry, . . .	[illegible]	71	[illegible]	[illegible]	[illegible]	[illegible]	[illegible]	[illegible]	74	[illegible]	[illegible]	.	[illegible]	.	[illegible]	.
Longford, . . .	[illegible]	[illegible]	[illegible]	[illegible]	[illegible]	[illegible]	[illegible]	.	.	[illegible]	.	.	.	.	.	.
Louth and Drogheda, County of Town.	73	[illegible]	[illegible]	[illegible]	74	[illegible]	[illegible]	.	.	.	.	.	.	.	.	.
Mayo, . . .	73	[illegible]	[illegible]	[illegible]	.	[illegible]	[illegible]	.	.	[illegible]	.	.	.	.	.	.
Meath, . . .	[illegible]	[illegible]	[illegible]	[illegible]	[illegible]	[illegible]	[illegible]	.	.	[illegible]	.	.	.	.	.	.
Monaghan, . . .	[illegible]	[illegible]	[illegible]	[illegible]	.	[illegible]	[illegible]	.	.	.	.	.	.	.	.	.
Queen's, . . .	[illegible]	71	.	[illegible]	.	[illegible]	[illegible]	.	.	.	.	.	.	.	.	.
Roscommon, . . .	[illegible]	[illegible]	[illegible]	[illegible]	[illegible]	[illegible]	[illegible]	.	.	[illegible]	.	.	.	.	.	.
Sligo, . . .	[illegible]	[illegible]	[illegible]	[illegible]	[illegible]	[illegible]	[illegible]	.	.	.	.	.	.	.	.	.
Tipperary, . . .	[illegible]	[illegible]	[illegible]	[illegible]	[illegible]	[illegible]	61	.	.	[illegible]	.	[illegible]	[illegible]	.	.	.
Tyrone, . . .	[illegible]	[illegible]	[illegible]	[illegible]	[illegible]	[illegible]	[illegible]	.	[illegible]	[illegible]	.	.	[illegible]	.	.	.
Waterford, . . .	[illegible]	[illegible]	.	[illegible]	.	[illegible]	[illegible]	.	.	.	.	.	.	.	.	.
Westmeath, . . .	[illegible]	[illegible]	[illegible]	[illegible]	.	[illegible]	[illegible]	.	.	[illegible]	.	.	.	.	.	.
Wexford, . . .	[illegible]	[illegible]	[illegible]	[illegible]	.	[illegible]	[illegible]	.	.	.	.	.	.	.	.	.
Wicklow, . . .	[illegible]	[illegible]	.	[illegible]	[illegible]	[illegible]	[illegible]	.	.	[illegible]	.	.	.	.	.	.

OBSERVATIONS

OF THE

DISTRICT INSPECTORS OF THE ROYAL IRISH CONSTABULARY AND OF THE SERGEANTS OF THE METROPOLITAN POLICE,

WHO ACTED AS SUPERINTENDENTS OF ENUMERATION;

IN REPLY TO A CIRCULAR DATED 29TH OCTOBER, 1879, ON THE PROBABLE CAUSES TO WHICH THE GOOD OR BAD YIELD OF THE VARIOUS CROPS IN EACH OF THEIR DISTRICTS MAY BE ATTRIBUTED.

PROVINCE OF LEINSTER.

PROVINCE OF LEINSTER.

CARLOW COUNTY. *Bagenalstown D.*—The yield of the various crops in this district, which are all good, is attributed to the favourable season. The corn was saved without much wet; the crops got rain at intervals in the early part of the season, which prevented farmers having to sow the seed in dry land, consequently giving an abundant return. The potato crop in this district is very good. *Carlow D.*—The grain crops in this county have been generally fair and well saved. Wheat, which was early sown, was much better than that which was sown later on. In dry and sandy soils oats was indifferent, but elsewhere good. Barley, a very fair crop, indeed. Potatoes good. Turnips &c., bad, owing to dry weather in the early part of the season.

DUBLIN COUNTY. *Balbriggan D.*—The average yield of the various crops this year being very good, I attribute same to the fine season. *Clondalkin D.*—Wheat and oats about same yield as last year. Barley, not so good as last year, owing to rain and damp at the time of ripening. Potatoes, not so good as last year, as large portion are blackening in consequence of the wet season. Turnips, mangold, cabbage, same as last year. Hay, better crop than last year, owing to wet season. *Colligmount D.*—The average rates of produce for this year in my district were nearly similar to those of 1878, there being a slight increase of the potato and vegetable crops, principally owing to the suitability of the season for the growth of the above mentioned green crops. *Dundrum D.*—The harvest generally has been fairly good. The potato crop, which promised well, seems to have suffered considerable injury from the recent rains. *Kingstown D.*—There is a slight increase in all crops in consequence of the more favourable weather during the season. *Lucan D.*—The crops generally throughout this district were fully up to the average yield both in quantity and quality. Turnips and mangolds were below the average in consequence of a run of dry weather in the early part of the season. *Rathmines D.*—There is a slight increase in the produce of the different crops, in consequence of the favourable season.

KILDARE COUNTY. *Athy D.*—The yield of the various crops is somewhat improved which, in my opinion, may be attributed to the absence of frost in the earlier part of the season, the farmers applying more manure, and a better system of cultivation having been established than in previous years. *Kildare D.*—The crops of all kinds are very good in this district this year. Hay crop, very heavy and well saved. Potatoes not yet dug out, owing to the heavy rains. *Naas D.*—The yield of the various crops may be considered good on the whole, and this, I believe, to be attributable mainly to the favourable weather while the various farming operations were going on. *Robertstown D.*—I consider the grain crop in general good, though a bit short of straw. Harvest weather was fine, so labour cheap. Potatoes in upland an average crop. In bogland a bad crop, as the early frosts in May, &c., greatly affected the tops. Root crop decidedly good. Turnips were down early, and then got the rain. I have never seen a better yield than the present. Hay good, and well saved. In general, crops decidedly above average.

KILKENNY COUNTY. *Callan D.*—The yield of the various crops this year is fair in comparison with years past, which may be attributed to an improved method of cultivation, and the good season which we have had. Turnips suffered much through want of rain. However, the yield compares favourably with that of previous seasons. *Castlecomer D.*—The yield of the various crops in this district, viz., wheat, oats, barley, and hay, may be considered good, and the potatoes were exceptionally good. Turnips and mangolds were rather poor. Carrots are not generally sown here, but the quantity was good, and the same may be said of the cabbage crop. Last spring was very favourable for farming operations, and the summer was very warm and dry. *Johnstown D.*—The various crops raised in this locality are in general very good, being considerably above the average in quantity and quality, which is to be attributed to the season weather during the harvest season, and the more than usual attention paid by the farmers to the ploughing, seeding, and weeding of their holdings. *Kilkenny D.*—The crops are generally good in the district. There was moisture enough to give a good crop of hay, but not sufficient to injure the corn. Roots also have had sufficient rain to make them better than for some few years past. Potatoes are a very fine crop and cheap enough to be obtainable by the peasant. *Piltown D.*—The past season has been most favourable for the growth and ripening of the various crops. No crop has failed; all are up to, or nearly up to, the average. The most successful was the hay crop, which was much above the average both in quantity and quality, a warm, moist spring favourable for the growth of grass being succeeded by dry, sunny weather at the season for haymaking. The corn crops are above the average, and were saved in splendid harvest weather. Root crops are about the average. Potatoes are a good crop, about the same yield as last year, but there is more disease, owing to heavy rains before and after the corn harvest. *Thomastown D.*—Wheat and barley crops, which are the most important in this district, were above the average this year. Oat crop also yielded well. Potatoes also good, with little sign of disease. Hay crop excellent, and well saved, the weather being very favourable. The only crop which may be put down as a failure was the turnip crop. This was owing to want of rain for a long period after being sown.

PROVINCE OF LEINSTER.

been; a very fair crop. *Mullingar D.*—The yield is about average.

WEXFORD COUNTY. *Enniscorthy D.*—This has been, by all accounts, the best farmers' year for a long time; but their hopes as to the value are very [illegible]. The hay crop was so heavy that the [illegible] was considered high. It was chiefly saved without a drop of rain. Barley was so abundant that the market became, and remains, glutted. [illegible] the top prices. The potato crop is various. I know two lands, not a mile apart. In one case the crop is large and abundant, in the other it is a total failure. The seed was bought from the same man at the same time, and was the same seed. [illegible] opinion that the ground, in many places, has been quite destroyed by artificial manure. Turnips promised to be a failure, and were resown in many lands. Rain, however, came, and the crop is now an average one. *Gorey D.*—The crops were fairly good, owing chiefly to the fine weather. *New Ross D.*—This has been an exceptionally good year for all crops usually raised in this county, except turnips, the latter in some parts of this district having failed in consequence of the time when the seed was sown being very dry; however, in most places the crop has attained the average produce. I believe that this year more land was under crops than for any ten years past; and although little manure was used, the cereal and other crops surpassed the farmer's expectations. There is abundance of hay in the district. Barley was very successful; and although the price is low, yet the good produce fully compensates for all. Potatoes were very successful, and no complaints about them are made in any way. *Wexford D.*—The yield of the various crops in this district for present year has been good, with the exception of the barley and turnip crops. The [illegible] of the former not being [illegible], was owing to the latter end of the spring being so wet, the time for sowing it; and the latter failing owing to the dryness of the month of June.

WICKLOW COUNTY. *Bray D.*—The cause of so favourable a yield of crops this year is owing to mild fine weather in the spring, followed by a good summer to ripen, and a most favourable autumn to gather in such. *Dunlavin D.*—Generally the good yield of all crops in this district this year is attributable to a mild spring, and favourable harvest weather. *Tinahely D.*—On the whole, the produce seems to be slightly under average. Oats and turnips are not so good as last year. It is difficult to account for the falling off in oats. The harvest weather was fairly good, and good crops were anticipated. There is also a falling off in turnips; the yields of flax [illegible], as little is grown in this district. The hay crop, too, was well up to average; but hay on old meadow land was inferior, the weather of the early summer being too dry for it. Other crops were above average. *Wicklow D.*—The produce this year, I am informed, is over the average, caused by the fine dry season. The turnip crop has suffered slightly from some cause.

PROVINCE OF MUNSTER.

PROVINCE OF MUNSTER.

CLARE COUNTY. *Ballyvaughan D.*—The produce of lands in this district this year appears to be over the average as regards all crops. This is to be attributed to the exceptionally fine summer and harvest. *Ennis D.*—The hay crop is plentiful, but not so abundant as it promised to be, owing to heavy rains at the latter end of summer; though the cereal crops (of which in these parts a comparatively small quantity is sown) proved up to the average. The root crop, owing to a suitable season, has turned out to be very satisfactory. The potatoes are plentiful and sound; the weather being propitious. *Ennistymon D.*—The yield in hay, turnips, and mangolds has been considered good, owing to the season favouring the growth and maturing of these kinds of crops. Cabbage has also been considered a good yield, and farmers who planted this vegetable for feeding purposes for cattle are high in its praise. The potato crop, which is the most important of all, is much and generally complained of, owing to the continual failure of the champion, which in this locality has ceased to be the productive crop which in previous years it proved to be. I, in a great measure, attribute this degeneracy to the general system of tillage persevered with in this locality, that is, continually tilling the one field, and with the same crop, which in such cases is forced on by artificial manures or seaweed. Bad tilling and no subsequent weeding are [illegible] instead of one crop the farmers here usually have about fifty different ones. *Kildysart D.*—The good average yield of various crops is due to the seasonable weather of [illegible]. *Killaloe D.*—The crops in this district, generally speaking, were quite up to the average this year, except the potato crop, which in some localities suffered from the unusually long spell of dry weather in the early part of the season. All the other crops were abundant and of fairly good quality. *Kilrush D.*—The crops this year were good above the average, the rain came in the right time for the hay, which is the staple produce of West Clare, little cereal crop being grown here; the potatoes are a good crop, both as to quantity and quality. *Scariff D.*—Hay was a good crop in this neighbourhood, wheat also, but oats and potatoes were not nearly so good as usual. May and June were very unfavourable, and it is probably owing to this that the various crops were not better. *Tulla D.*—The difference in the good or bad yield of the different crops in the several electoral divisions was caused by the difference of the land. In some of the divisions the land is very bad and mountainous, while in others it is very good. As a whole, the crops in general were very fair.

CORK COUNTY, E.R. *[illegible] D.*—The crops all round were good in this district, owing to the good season, and the fine weather during the autumn for saving and stacking them. *Castletown D.*—This was an exceptionally good year. The yield of the various crops is rather above the average, and this is believed to have been caused by a favourable season, the weather being all that could be desired by the farmers. *Cork, North, D.*—The crops in this district have been very good this year, particularly the potatoes and hay crop. Root crops were tolerably good, notwithstanding the hot weather in July last. *Cork, South, D.*—The crops on the whole for this season have given a very good yield, which may be attributed mainly to a course of good weather. *Fermoy D.*—From statistics which I have collected, it would appear that in the greater part of this district the hay and oat crops are better than, and the potato crop is not so good as, in 1858. On the whole the crops were good, owing to the favourable weather, especially the absence of much rain during harvest. *Kinsale D.*—The harvest this year in this district is one above the average, every crop yielding abundantly. Prices are fairly high and remunerative, and I attribute this to a better feeling existing in the country, whereby trade has improved, and this, taken in conjunction with an abundant harvest, has considerably improved the position of the tenant farmers. *Macroom D.*—With reference to the rates of produce of the several kinds of crops in the electoral divisions of this district, which on the whole (turnips excepted) were good, I beg to state that the cause attributable to such, in my opinion, is that the weather was so favourable during the season, the cereal crops were allowed to fill and come to perfection, whereas in other years past the heavy rain in the

[illegible]

district may be attributed to the wetness of the season. *Tralee D.*—The season was favourable for all crops in this district, save that of the hay crop. Owing to the continuous drought in midsummer the hay crop is not as good as that of last year.

LIMERICK COUNTY. *Abbeyfeale D.*—The general produce of all kinds of crops has been good in this district this year. This is attributable to the favourable spring, the [illegible] summer, and the fairly good harvest. Most of this district is mountainous and, as a rule, rain prevails. This year the spring was dry, which enabled the farmers to cultivate their lands in good time; the summer was dry, and there was sufficient dry weather in harvest to enable all to secure their crops. *Adare D.*—The year was generally favourable as regards all crops. There is no extraordinary yield in any one of them, nor is there any marked falling off. The land in this district is much better in some parts than in others, which will account for the great difference shown in the yield of potatoes in some places. *Bruff D.*—The greater portion of the land in this district is under grass, as it is chiefly used for fattening cattle and for grazing milch cows. Such small quantities of wheat, oats, and barley as were sown, yielded very fairly, but the wet weather injured those crops to some extent. Potatoes yielded well, but a considerable portion of them, in some localities, became diseased, specially "champions." I think the heavy fall of rain before the potatoes were fully matured, contributed to the spread of disease amongst them. The rain which affected the potatoes injuriously, had a contrary effect on turnips and mangolds, which yielded very well. Hay is a good crop, but a considerable quantity (specially in low land) was not well saved, owing to the fall of rain. There appears to be an abundant supply of hay in the district, and there has not been much of it consumed up to the present, except by horses, as the weather latterly has been mild, and there is grass to be had in the fields for cattle. *Kilfinane D.*—Owing to the summer months being so favourable all crops grew well in this district. Wheat and oats were a very fair crop. The potato crop was good, and all green crops, such as turnips, mangolds, cabbages, &c., were good. The hay crop was very good. *Limerick D.*—As regards the yield of crops this year, mangolds, turnips, and barley have been below the average, this is generally attributed to the wetness of the season. The other crops of corn and hay have been very good, hay exceptionally so, owing to the season favouring them. *Newcastle D.*—The yield of the various crops in this district for the present season, is a fair average. The proportion of land under tillage is gradually becoming smaller, owing to the present high price of cattle. The district, however, has always been mainly a grazing one. Hay and grain crops are good, and as the harvest weather was favourable, have been saved and stored at small expense. Green crops, with the exception of potatoes, are not up to the average of former years. This may be attributed to the long period of dry weather in the months of May and June. The potato crop is good, and not largely affected by disease. *New Pallas D.*—The increased produce in some crops, particularly wheat and oats, is attributable to the favourable condition of the land when planting, the months of March and April being dry, the subsequent rains in May, followed by favourable weather in August and September, for ripening and saving. A slight decrease is noticeable in the hay crop from 1858, but it is considered a good average crop. Potatoes, turnips, and mangolds are also an average crop this year, but the two latter are not grown to any great extent in this district. *Rathkeale D.*—Wheat is a very good productive crop this year, which may be attributed to the very mild, favourable winter of 1858, just after it had been sown. Oats and barley are not so good, and the yield of each crop is not so abundant as that of wheat, owing to the month of June setting in so dry, which proved very deleterious to these two cereals. Potatoes are a good, abundant crop, but many complaints are being made of the disease in this crop, which is mainly attributed to too much rain having fallen in the month of August, just as the potato was nearly ripe, and had almost ceased growing. Some farmers allege that one-half of this crop is "black," or diseased, and others say that one-third, at least, is thus [illegible], but the yield is a fair one, and quite sufficient to meet the wants of the people whose principal support is derived from the potato. Turnips, in general, are good, but some farmers look upon it as a poor crop this year, on account of the dryness of the season at the very time when the seed was put into the ground, and for some short time afterwards, as vegetation was about to set in. The yield in other green crops is considered very fair, which may be ascribed to the weather, which turned out favourable to their growth. The hay crop, although of good quality, has not been quite as abundant as in previous years, owing to the severe dry weather in early summer, which proved very injurious to the yield in this crop, particularly in upland. The yield was not so heavy as in previous years, but the quality of the hay was much better, and the crop was secured in good time by the farmers.

TIPPERARY COUNTY, N.R. *Borrisokane D.*—The potatoes are not so good in this district as last year, owing to the blight having set in so early, and the turnips are not up to the average, owing to the dry season. All other crops are very good. *Nenagh D.*—Barley and oats, good crops, owing to the fine weather in June, July and August. Hay very good, owing to the rains of May and early June, and the warm weather of July and August. The dry weather also served the potatoes, but had a very bad effect on turnips, which, however, somewhat recovered themselves by the late rains. In most cases they had to be re-sown. *Newport D.*—The crops as a rule were very good throughout this district, owing to the mild spring and dry summer. The potato crop, however, suffered from the wet weather in spring, and is not so good as last year. *Roscrea D.*—The chief reason the crops were so good this year in this district was the fact of the exceedingly favourable weather we had. I also hear that more manure and better tilling were indulged in. The mangolds and turnips were the only poor crops, attributable to the weather not suiting them in the early stage of growth. *Templemore D.*—The yield of the various crops I find is considered very fair. Owing, however, to the dry season at the commencement, a good deal of the turnip crop has failed there [illegible], but those that did come on are generally a splendid crop, and almost counterbalanced the loss in those that failed. The good yield generally is owing to the favourable season, besides the land about here is good. *Thurles D.*—The crops generally are better than last year, which I attribute to the favourable season.

TIPPERARY COUNTY, S.R. *Cahir D.*—All the crops this year were very good. Above the average, owing to the weather being so suitable. *Carrick-on-Suir D.*—The slightly increased rate of produce in this neighbourhood appears to have been due to the favourable weather. The potato blight has not hitherto done much harm in this neighbourhood. *Cashel D.*—I believe that all round the crops of every sort have been exceptionally good this year. I believe that a favourable state of the weather at the different periods when it has such an effect on the prospects of the farmers, has been the cause. *Clonmel D.*—The average produce of wheat and oats in this district this season has been rather above that of former years. This may be attributed to the early part of the season being moist, and a good harvest. Meadow crops are also above the average, owing to same cause. The potato crop in this district is not quite so good. I think the early part of the season being wet was the cause of this. Turnips also are not quite up to the average, owing to the dryness of latter end of May and June.

Taking the crops generally in this district, the average yield all round is very satisfactory. *Dunmanway D.*—I beg to say the harvest has been abundant. The good season has been productive of very heavy crops. The cereal crops full in the grain and rich in the stalk. Potatoes, turnips, and other roots abundant and free from injury. The varying heat and rain throughout summer and a dry early autumn, precisely suited the growth and in-gathering of crops in this part of the country. *Killenaule D.*—I attribute the good harvest of all crops, corn, hay, &c., in this district this season to the fine weather we have had. *Tipperary D.*—Wheat, oats, and hay were a most excellent crop, owing to the favourable season. Turnips, mangolds, carrots, and cabbage were fair. Potatoes not so good as usual, owing to the drought in early part of season.

WATERFORD COUNTY. *Cappoquin D.*—The harvest this year was a very good average one. All crops were good, with the exception of turnips, which were injured by the dry weather of the early part of the season. The hay crop was very fine. *Dungarvan D.*—On the whole, there has been a good yield for all crops. There is a slight decrease in the potato, turnip, and carrot crops, which is supposed to be owing to the dry summer, in consequence of which the seed did not come up evenly. There is an increase in the produce of wheat, oats, and barley, which is attributed to the favourable harvest. The grain ripened properly, and was saved, and also produced a good yield. There is not much of an increase or decrease in other crops. *Portlaw D.*—The yield of all crops in my district has been good, and I think this is to be attributed to the exceptionally good weather of the spring and summer. *Waterford D.*—The yield of the various crops in this district, as compared with recent years, is good. This is attributable, chiefly, to the good weather that prevailed during the season for harvesting operations. Also, the farmers seem to give more attention to agricultural pursuits, and an increased spirit of industry on their part is setting in.

Province of Munster.

PROVINCE OF ULSTER.

Province of Ulster.

ANTRIM COUNTY. *Antrim D.*—Owing to the very dry weather in the early summer, the flax crop suffered severely; and it is, generally speaking, a light crop throughout this district. The grain crops, for like reasons, were rather lighter than usual. The other crops appear up to the average. *Ballymena D.*—Generally the crops have been very good in this district. Hay, of which a large quantity is used in this district, was an abundant crop and capitally saved, the weather being most suitable at the time. Corn also was a very capital yield, but suffered to a certain extent in ripening, owing to the very wet weather in July and August. Potatoes have given a very great yield, but they are not as dry a crop as might be wished, owing to a somewhat similar cause. Flax was a very good crop, the weather being specially suitable for it. Turnips were satisfactory in their yield, and little affected by fly. On the whole, this has been [illegible] above the average harvest. *Ballymoney D.*—Owing to the weather having been unusually favourable, the hay crop was above average produce, and better saved than for many years past. The various other crops in this district were about average yield. The potato crop, however, was somewhat injured by the heavy rains about [illegible], and, consequently, is not quite so good as it would otherwise have been. *Belfast, North, D.*—The yield of the grain crops this year was fair in quantity and quality, owing to the [illegible] coming on when it was ripening. As regards the potato crop, the yield has been good, but the quality was considerably damaged by rain, thereby bringing the crop an average one only. Hay has been an abundant crop, but the fact of the weather coming [illegible] and wet at the time of saving, greatly diminished the value of the crop. Respecting turnips, mangolds, and cabbage, the yield has, on the whole, been tolerably satisfactory, owing to the moist weather which in the [illegible] particular season) [illegible] favourable to their growth. *Belfast, South, D.*—The rates of produce are larger than in other years, the probable cause being the [illegible] weather. *Carrickfergus D.*—The crops, generally, this year are abundant, but not well saved owing to the [illegible] weather. Many farmers are "penny wise and a pound foolish"; by saving a pound or [illegible] in labour they risk the saving of their crops, and they do not pay in many instances the necessary attention to cleaning their land of weeds or [illegible] good seed. *Lisburn D.*—The yield this year is little, if anything, above the average; this I attribute to the exceeding hot weather during spring-time and up to the end, I may say, of August, which is [illegible] state of things so far. Potatoes in this district have been exceptionally good. Oats of fair yield, but short in straw.

ARMAGH COUNTY. *Armagh D.*—The small yield this year is attributed to the early part of the season being so dry that the crops were not got in early enough; then the weather got so dry that the land became hard, and to a great extent prohibited growth of all crops, particularly potatoes, turnips, and mangolds. *Lurgan D.*—Hay was a good heavy crop, but that some of it was much injured by the wet weather in the latter part of the season. The [illegible] was light owing to the dryness of the ground when the seed was being put in; but the crop improved with the subsequent rain. The same may be said of the other root crops. The grain crop was much injured by the continued wet season, and was in many places prevented by it from coming to maturity. *Newry D.*—Taken all round crops were a good average. *Portadown D.*—The good crop of early hay was, in my opinion, attributable to the rain in early spring. The potato crop was not good, in consequence of the continued drought until late in the season. The barley crop and spring wheat were injured by same cause. The old meadow hay was injured by the rain in August and September, and a large quantity along the banks of the River Bann was flooded, and suffered much damage. Taking the crops generally in this district a very fair average is to be. The farmers are receiving good prices for the produce of their farms.

CAVAN COUNTY. *Bailieborough D.*—On the whole this season has been a favourable one. The potatoes were very good and completely free of blight. The yield of oats was good, but the straw was short. The hay crop was abundant, exceedingly good and well saved. Turnips were light. The flax crop was middling. The dry summer was favourable for the potatoes, hay, and oats, but there was not enough moisture for flax and turnips. *Ballyjamesduff D.*—The crops in this district were generally good this season. Hay was good, and well saved, owing to the dry weather in June and July the crop was not so heavy as in 1888; oats is the principal corn crop, and it is very fair; the dry summer caused the straw to be short, but the yield on the whole is very good. Potatoes are good for food owing to the dry warm summer, and the yield is very good, but they are not so large as in 1888; the crop came to maturity early this year, and being good there appears much more of the crop used at this season than there was of last year's crop at this time of the year. Turnips are fair, but hardly so good as last year owing to the dry summer; cabbage is generally a good crop. Flax, there is little grown of this crop in this district, the same applies to wheat, barley, &c. Turf, for fuel, is plentiful as the summer

PROVINCE OF ULSTER.

was suited to drying of it. *Clones D.*—The crops were generally good this season; the seed time was dry, and the ground in good order for sowing. *Killeshandra D.*—The potato crop was considerably injured, especially on light soils, by the long drought of June and July. Hay also suffered from the same cause, and so did the turnips and flax. The other crops appear to be up to the average. *Bawnboy D.*—The potato crop is not so good as anticipated; except in very good and dry seasons it seldom reaches a good average, on account of the humid nature of the soil. The hay crop is fair, as the season for saving it was remarkably favourable. The yield of grain crops is on the whole a fair one. Root crops are cultivated to a limited extent only, but the yield is considered on the whole an average one. The yield on the whole may be considered fair.

DONEGAL COUNTY. *Ardara D.*—There is no marked difference in the yield in most crops. Potatoes are a good crop, much above the average. This I consider principally due to a favourable season. The other crops grown in this locality are about an average, and do not therefore require any observation. *Ballyshannon D.*—Crops are all up to the average in this locality. Potatoes are a good crop, but would have been better but for the heavy rains in the early part of August. Turnips are above the average. Early meadows were suited by the fine weather in June and July. Straw, which at one time appeared likely to be a short crop, is up to the average. *Buncrana D.*—There is a fairly good yield of crops, which is attributable to fine weather in the early summer. The yield in cereals would have been better, but for the extreme wet in the later summer and harvest. *Dunfanaghy D.*—I consider that the favourable yield this year is due, in a great measure, to the mildness of the season in comparison to others. The growth was considerably checked at the commencement of the season, owing to continued dry weather, but the rain setting in afterwards helped to facilitate considerable progress in both growth and yield before a continuance of one class of weather had time to work its effect to any extent. *Dungloe D.*—On the whole, the harvest of this year has been good. The potato crop was very productive. This crop is regarded as the most important one by the inhabitants of this district, as in it they have their chief means of subsistence. The grain and hay crop have also been fully up to the average, probably as good as they could have been when the nature of the soil is taken into consideration. *Letterkenny D.*—The potato crop, though perhaps a little above the average of last year, is far under what was expected in the early part of the season, owing to the wet weather, which has done so much harm. The oats is about the average of last year. Turnips and flax is about the average of last year. Hay, in many places, has been greatly injured by wet weather, and is, I believe, under the average of last year in this district. *Moville D.*—All the crops throughout this district were good. A good yield of potatoes. The corn crop was also above the usual yield. The season was very favourable on the whole, to which that the good yield must be attributed. *Raphoe D.*—The potato crop in this district, generally speaking, was up to the average yield, except in marshy places. This is attributed to the moisture early in the season and the warm weather succeeding in June and July. The turnip crop was good, and owing to the same reason. The flax crop was better than last year, and the people about here speak of it as much better than the usual yield. The oat crop, early in the season, promised better than it ultimately yielded. This is due to the lack of sufficient moisture in the summer months, which [illegible] made the cereal crops unusually stunted.

DOWN COUNTY. *Banbridge D.*—The potato crop is much below the average of late years, owing, chiefly, to the wet spring and the drought in the months of June and July. The hay crop is above an average, owing to wet weather in May. Oats is below the average, owing to the cause of failure in the potato crop. Flax, turnips, mangel, wheat, &c., were greatly retarded in growth, owing to the very dry weather in June and July. *Downpatrick D.*—The yield of the several crops in this district is good. I attribute this to the favourable weather of the past summer. *Newtownards D.*—There is a fair average yield of the various crops grown in this district, with the exception of flax, oats, and turnips, sown in dry sandy lands. The falling off in the yield from the three last-named crops is attributed to want of rain in the early summer. Flax, oats, and turnips, grown in damp soil are a good average. Wheat is a fair average crop, and hay and potatoes are exceedingly plentiful. *Rathfriland D.*—The weather, which was dry in the late summer and wet in early part of harvest, appears to be the principal or only cause affecting the yield of the various crops. I am not in a position to contrast the yield with that of former years.

FERMANAGH COUNTY. *Derrygonnelly D.*—Neither the yield of the potato or oat crop are as good as that of 1879. The land in this district is naturally of a stiff and heavy nature, owing to the quantity of fresh water in the locality. The spring season was very wet, and both oats and potatoes were late in sowing in consequence. The harvest was also very wet, and the potato crop was attacked with disease before the tubers were matured. Turnips is an average crop, as good as last year. Hay is a good crop; but a good deal of it has been badly saved owing to the continued wet in the harvest time. *Enniskillen D.*—The various crops in this district, with the exception of potatoes and turnips, are a fair average of former years. The potato crop is scarcely up to the average, owing to the blight having set in early in the month of August, coupled with an unusual amount of rain in same month. The turnip crop suffered in consequence of the dry weather in the months of May and June. *Irvinestown D.*—Owing to the wetness of the early spring in this district, the crops were not put in until rather late, and this fact, combined with the dryness of June and the early part of July, which is unfavourable to light soil, accounts for the rather light yield of grain and root crops. The same causes were highly favourable to the hay crop, which was very good, and for the most part well saved. In the low-lying lands of the district, the turnip crop was good, but it was an utter failure in the high ground, as the drought, followed by heavy rain, prevented its being sown in good time. *Lisnaskea D.*—In this district the crops have been fully up to and above the average. Hay cheap and plentiful. Potatoes on damp ground above the average, on dry ground under the average. On the whole the potato crop well up to average. Turnips and mangold sown late were, in many cases, a failure; but where they grew were a good crop. Oats were poor, thin and light, and both head and straw deficient, owing to dry spring. Flax was fair in quality, but few really good samples have been seen; in general it was short in staple and light. Wheat is but little grown here, and the little I saw appeared good in quality and quantity. On the whole an excellent year, above the last two.

LONDONDERRY COUNTY. *Coleraine D.*—The yield of the various crops in this district has been generally a good average, not calling for special remark. *Limavady D.*—The yield of the various crops generally considered has been above the average in this district. Potatoes and turnips especially are an excellent crop; rain having fallen when both most required it, and so developed them. Hay and straw are of excellent quality, but rather short, owing to the drought in the early part of the season. Oats gave an average crop, but would have been better but for the heavy rains just previous to harvest time. Flax is good, but rather coarse for the same reason. *Londonderry D.*—The crops in general this year are good. This is due in a great measure to the early summer being wet, and the month of June being very warm and dry. The autumn too has been most favourable, especially to green crops, which are still growing. The potatoes are a plentiful crop, but the early kinds suffered considerably from the autumnal rains, and many rotted in the ground. The rural district being small, I cannot consider these returns afford a correct standard by which to judge. *Magherafelt D.*—The crops generally are a fair average this season. Were it not for the long and excessive drought in the early part of the season, the potato, oat and turnip crops, would have been, in all probability, much better.

MONAGHAN COUNTY. *Carrickmacross D.*—The hay crop was a [illegible] heavy crop this season [illegible] [illegible] of the [illegible] [illegible]. But the turnip crop was [illegible], owing to the drought at the time of putting in the seed. *Clones D.*—The farmers of this district have extensively [illegible] [illegible] of Lord [illegible] Act in the purchase of their farms. Having become landed proprietors they devote themselves more zealously to agricultural pursuits. This fact, coupled with a favourable season, accounts to a great extent for a [illegible] than average yield of various crops. *Monaghan D.*—The yield of the several crops in this district appears to be fairly good. In the electoral divisions of Tullycorbet, Rockwallace, Crossmoyle, Annayalla, Shankill, [illegible], and Clones the land is of an inferior quality, and the farms are small and not so well cultivated, which accounts for the crops not being so good.

TYRONE COUNTY. *Aughnacloy D.*—The variations in the yield of the several crops are due mainly to climatic influences. The farming in this district is generally good and cleanly, but not of a very high class, the greatest fault being shallow tillage and not being sufficiently mindful of manurial products. *Clogher D.*—All the crops were up to a fair average, except flax and potatoes. The long drought in the beginning of the season, and the excessive rain towards the end, did a great deal of damage to these two crops; they are both much below the average and of bad quality. *Dungannon D.*—The yield of the various crops appears to have been better than the average this year, and this is obviously attributable to the favourable weather. *Newtownstewart D.*—I beg to state that as the system of agriculture in this district is almost the same each year, the yield of the various crops (whether good or bad) is attributable to the favourable or unfavourable state of the season. *Omagh D.*—I am of opinion that the crops produced in such fine abundance were the result of the good weather. *Strabane D.*—The yield of the various crops in this district may be considered a very fair average, and that there is nothing calling for any special remarks regarding the yield.

PROVINCE OF CONNAUGHT.

GALWAY COUNTY. *Athenry D.*—The crops were a fair average in this district this year, and there has been no complaint by the farmers so far as I am aware. In some parts of the district, oats, potatoes, wheat, and hay were better than others, but on the whole there is no reason for complaint. Some farmers manure and labour better than others, hence better crops. The season was favourable, to which the good yield may be attributed. *Ballinasloe D.*—In this district this has been a very good average year. Potatoes are small, owing to the drought in the early part of the season, and are getting black in a good many instances in consequence of the late heavy rainfall. Oats and hay are very good, owing to the favourable season, and mangolds and turnips are very good also. *Clifden D.*—The yield of crops in this district for the year has not been as good as last year, which I attribute to the wet weather. The [illegible] [illegible] has been good, and [illegible] [illegible]; but the potato crop is not good, [illegible] [illegible] [illegible] crop. Potatoes on dry sandy land in some parts of the district have done very badly, while on boggy land they have done rather better. *Gort D.*—I attribute the comparatively good [illegible] yield of the several crops in this district to the fine weather experienced during last spring, when the crops were all got into the ground dry. The early [illegible] of summer was more favourable for the growth of [illegible] than [illegible] crops. The land in this district being mainly limestone, a dry spring and summer favourably affects the growth of the crops planted in the district, which are mainly potatoes and oats. *Loughrea D.*—It seems that the oat crop this year was fairly good, owing to its having been favoured by good weather at the right time. The potato crop, the staple one in this district, is, I am informed, not so good as was expected. The reason is ascribed to too little wet weather at certain periods. *Galway D.*—A great portion of this district is of limestone formation, and the dry weather, early in the season, had an injurious effect on most of the crops. For the same reason the crops in low-lying and boggy lands were good. On the limestone land the potato, turnip, mangold, and oat crops were injured by the dry weather early in the season. The wheat, and rye, and barley crops were good, as the dry weather benefited them. In many places the potato crop was seriously affected by the blight. As a rule, early-sown root crops were good. There was a fair crop of hay. On the whole, the harvest was an average one. *Gort D.*—The light yield in the oat and wheat crops in this part of the country appears to have been due to the unseasonability of the season, as also in that of hay. Root crops, such as turnips and potatoes, were abundant, the season being more suitable for them. *Loughrea D.*—The crops, on the whole, throughout this district, are of a fair average, save the potatoes, which are not quite up to the average of last year's crop. This falling off may, I think, be attributed to the dry summer, and the soil of the district around being of a limestone nature. *Mountbellew D.*—I am informed that the yield of oats and hay this year are fairly up to last year's average. The potato crop is not as good as last year's, and the deficiency was caused by frost in the beginning of July, and the heavy rain in August. The turnip crop is considered better than last year's, and this is attributable to the rain in August. The above are the principal crops raised in this neighbourhood. *Portumna D.*—I beg to report that the crops were up to the average this season, and call for no special comment. *Oughterard D.*—The insufficient yield of crops in this locality is due to too much rain and a poor stony soil. The decrease in the yield of hay and oats this summer was because there was more sunshine and less rain than is usual in these parts. *Tuam [illegible] D.*—There was a good yield of various crops in this district this year. The potato crop, however, was not as good as that of previous years. This is owing to the fact that the land was not so well manured this year as in others. Any falling off there may have been in the various crops is due to the heavy rains which fell in the autumn months. *Spiddle D.*—Throughout this district the crops, with the exception of potatoes, have been very good. The yield of potatoes is about 20 per cent. less than last year, probably owing to absence of rain in the early part of the

PROVINCE OF CONNAUGHT.

season. It will be observed that the average produce per acre in the electoral divisions of Gorumna Island, Lettermore Island, and Crumpaun is much higher than elsewhere, and this I attribute to the greater quantities of seaweed spread on the land by the inhabitants of said divisions. *Tuam D.*—The hay and wheat crops were good this year, which is attributable to the fine weather that prevailed during the summer. Oats and potatoes are average crops, and turnips and mangolds fair. On the whole, the farmers are well satisfied with the season. *Woodford D.*—As the season was not unfavourable, the average yield of the various crops in this district may, on the whole, be regarded as fairly good. In those electoral divisions where the average rates of produce per acre are returned as low the land is of a poor quality.

LEITRIM COUNTY. *Ballinamore D.*—The rates of produce of the various crops are generally good in this district. There is no wheat sown in this district. The yield of the oat crop is fairly good, but this crop does not ripen early here, and the continuous wet weather in latter part of September and beginning of October tended very much to retard its progress towards maturity, some of it had to be cut green, otherwise the yield would have been much better. The potato crop is a good one, owing principally to the dry summer which suits the description of the soil about here; there is, however, a considerable quantity diseased in some old lowland. The turnip crop is fairly good considering the nature of the soil. Cabbage is a good crop, the land selected for this crop being the best adapted for it. The yield of hay is fairly good, but not so abundant as last year. Any cut early in the season is well saved, but any not cut till end of September suffered very much from the heavy rain and floods. On the whole the farmers are well satisfied with the crops in this district. *Carrick-on-Shannon D.*—This harvest has on the whole been good and above average. The hay crop was very good and unusually early, and fully two-thirds of it was saved in perfect condition, one-third principally old lake meadows was more or less damaged. Oats, which is the only corn crop grown in the district, except a little rye, was an excellent crop, much over average as regards oats, and fully average as regards straw; most of it was also well saved. Potatoes are a fully average crop, and but for the season being rather wet towards the end would have been an exceptionally good crop. The wet weather, however, caused a good deal of disease and rottenness in the tubers. Very little green crops are sown in the district. Small patches of turnips, mangolds, and cabbage being planted here and there; anywhere sown they were all fairly good crops. Mangolds and turnips are on the small side owing to too much rain and want of sunshine. The general prospects of the farmers are fair enough. Cattle are very dear, though there is a slight fall in prices lately in every class, except half-year old calves. Butter is if anything dearer than other years, though the price is still too low. There is an abundance of food both for the people and animals, and I think there is no scarcity of fuel. *Drumshanbo D.*—Taken all round a fair yield of crops has been given in this district. The potato crop is not as good as was expected, owing to excessive rain. The crop of oats is good, as would have been hay, but for heavy rain during time of saving. The root crops, mangel wurzel and turnips, have been perhaps more satisfactory than other crops. *Manorhamilton D.*—The good or bad yield of the various crops may be attributed to various causes, viz.:—the manner in which the good or bad tenant manages his farm. Where the tenant attends to his business and cultivates his farm well, a good crop follows, and *vice versa* with the indolent man. This year the season being comparatively a dry one has contributed to a fair produce in all kinds of crops, the soil in this district being of a moist boggy nature. *Mohill D.*—The crops are fairly good everywhere this year which, may be attributed to the good season.

MAYO COUNTY. *Ballina D.*—On the whole the crops grown in this district are, I think, up to a fair average. Potatoes, one of the principal crops, are, however, scarcely so good as they were last year, this is caused by the very wet season we have had during the time they were coming to maturity, the land being generally cold and wet. Oats are fully up to the average of recent years; so are turnips; and as for barley, rye, &c., they are not extensively grown in this district. On the whole, farmers appear fairly well satisfied with this year's harvest. *Ballinrobe D.*—I think the average yield fair. Potatoes have suffered very much from the rain. *Belmullet D.*—The rates of produce of the crops, that are raised in this district, are more than an average for this part of the country. The good yield of the crops is due to a fairly dry summer. *Castlebar D.*—There is nothing calling for remark in the yield of the crops in general, but potatoes yield somewhat less owing to the wet weather at the time they were maturing. Hay is rather under the average, although the yield of grass was good, the wet weather made it difficult to save, and some of it was lost, and the quality is inferior. *Claremorris D.*—The wetness of the summer injuriously affected low-lying and swampy land, especially in regard to the potato crop, which is not up to the average this year. The other crops appear to be a fair average. *Newport D.*—I believe the crops in the several electoral divisions in this district are fairly good this year, which I attribute to the favourable season—May being rainy and June dry. *Swinford D.*—The crops in my district are on the whole good. The potato crop is not quite up to that of 1887, but I consider it better than last year. A good deal of damage was done to hay, oats, and potatoes in the valley of the Moy during the recent floods. *Westport D.*—The crops of this district for past year were middling.

ROSCOMMON COUNTY. *Athlone D.*—The potato, oats, turnips, and hay crops, are more than up to the average this year. Owing to the dry summer there is an excellent return in the potato crop. The straw of the oats was a little short, but has fair yield. Hay is plentiful, and of good quality. Same applies to turnips. *Boyle D.*—I consider the yield of the various crops for present year in this district to be, on an average, very good. This I attribute to the favourable season. *Castlerea D.*—The potato crop about a quarter or one-fifth less than last year. The excessive rain in June, July, and August, affected the growth, and caused the difference in the yield. The turnip crop is about the same as last season. The yield of grain is about the same as last year; but the straw is longer and better, owing to the wet season. The extreme rainy weather during the summer months caused the hay crop to be somewhat better than last year. *Roscommon D.*—The harvest here has been good. Owing to a dry, early summer, hay, although light, is plentiful. Oats, for same reason, is good; and the potatoes are above the average. *Strokestown D.*—The principal crops grown in this locality are hay, potatoes, oats, and turnips. Hay is a fair average crop. Potatoes is about one-third under the average, owing to having been attacked with blight and frost very early in the season. Turnip is a fairly good crop. Oats, a fair average crop.

SLIGO COUNTY. *Ballymote D.*—The yield of potatoes is somewhat less than that of last year, owing to the very wet season we have had this year. The grain crop is fair, and the hay crop is very good so far as the yield is concerned; but I am afraid it is not well saved, owing to this wet season. The yield of all crops in this district is fair. *Collooney D.*—The yield of all crops this season has been very large, and the crops sound and healthy. The hay crop especially is extremely large and good. This satisfactory state of things is due to the favourable climatic conditions which existed during the year. *Boyle D.*—Oats: good crop, would have been better were it not that the [illegible] was too dry. Wheat: the same remarks apply, and there is [illegible] wheat here. Barley: The same remarks apply as to wheat and oats. Rye: ditto. Hay: a good crop—favourable season. Potatoes: ditto. Turnips, good. Mangolds, good. Cabbages good. *Sligo D.*—There has been a good yield of all crops in all the divisions of this district, except oats or bere, which may be attributed to good soil and favourable weather. In the places where the yield was light, it was on account of bad soil. *Tubbercurry D.*—The principal crops grown in this district are potatoes, oats, and hay. Potatoes are an average crop, especially in good land, and escaped blight and fairly matured. In [illegible] districts not so good, owing to rain. Oats turned out a fair crop in low land. The rains of May were in its favour, and August was a good ripening month. Hay has been a plentiful crop, and fairly saved, especially that cut early, due to favourable weather. Turnips and other green crops, are very good, owing to a favourable season.

PROVINCE OF CONNAUGHT.

APPENDIX.—SILOS

The following statements have been received from persons who have made Ensilage in Ireland in [illegible].

SILOS AND ENSILAGE

AND ENSILAGE.

[illegible]	[illegible]	Walls	Floor	Roof	[illegible]	[illegible]	[illegible]
	[illegible]	[illegible]	[illegible]	[illegible]	[illegible]	[illegible]	[illegible]
	—	—	—	—	—	—	[illegible]
	[illegible]	[illegible]	[illegible]	[illegible]	[illegible]	[illegible]	[illegible]
	—	—	—	—	—	—	[illegible]
	—	—	—	—	—	—	[illegible]
	[illegible]	[illegible]	[illegible]	[illegible]	[illegible]	[illegible]	[illegible]
	[illegible]	Do.	Do.	Do.	Do.	Do.	
	[illegible]	[illegible]	[illegible]	[illegible]	[illegible]	[illegible]	
	[illegible]	[illegible]	Concrete	[illegible]	[illegible]	None	[illegible]
	[illegible]	[illegible]	Concrete	[illegible]	[illegible]	[illegible]	[illegible]
[illegible]	[illegible]	[illegible]	[illegible]	[illegible]	[illegible]	[illegible]	[illegible]
[illegible]	—	—	—	—	—	—	[illegible]

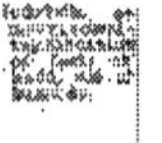

LEINSTER—continued.

[illegible]	[illegible]	[illegible]
[illegible]	[illegible]	[illegible]
[illegible]	[illegible]	[illegible]
[illegible]	[illegible]	—
—	—	—
—	[illegible]	—

—	—	—	—	—
—	—	—	—	—
[illegible]	[illegible]	[illegible]	[illegible]	Do.
[illegible]	Do.	Do.	Do.	Do.
[illegible]	Do.	Do.	Do.	Do.
—	—	—	—	
[illegible]	Stone and [illegible]	—	[illegible]	[illegible]
[illegible]	—	—	—	
[illegible]	Masonry.	[illegible]	[illegible]	Do.
—	—	—	—	—
—	—	—	—	—
[illegible]	[illegible]	[illegible]	[illegible]	Do.
[illegible]	Concrete.	Concrete.	[illegible]	Do.
[illegible]	Do.	Do.	Do.	Do.
[illegible]	Do.	Do.	Do.	Do.
[illegible]	—	—	—	—

LEICESTER—continued.

[illegible]	[illegible]	Temperature.	
		[illegible]	[illegible]
...	...	...	—
...	...	...	—
[illegible]	[illegible]	[illegible]	—
...	...	...	—
[illegible]	[illegible]	—	[illegible]
[illegible]	[illegible]	—	[illegible]
...	[illegible]	...	—
[illegible]	[illegible]	...	—
[illegible]	[illegible]	[illegible]	...
[illegible]	[illegible]	[illegible]	[illegible]

Dimensions of Silo—Length, Breadth, Depth.	Materials of Silo.			Whether Divided or not.
	WALLS.	FLOOR.	ROOF.	
[illegible] feet by [illegible] feet, [illegible] feet deep.	—	—	—	Not, .
—	—	—	—	—
[illegible] feet by [illegible] feet, [illegible] feet deep.	Brick, . . .	Paved and Cemented.	Slates, . . .	Not, . .
[illegible] feet by [illegible] feet, [illegible] feet deep.	—	—	—	Divided, .
—	—	—	—	—

LEINSTER—continued.

[illegible]	[illegible]	Temperature.		[illegible]
		[illegible]	[illegible]	
[illegible]	[illegible]	About 60 Degrees.	Constant [illegible]	About 12 or 14 lbs.
[illegible]	[illegible]	[illegible] degrees.	[illegible] degrees.	[illegible]

PROVINCE OF

[illegible]	[illegible]	[illegible]	[illegible]	[illegible]			[illegible]	[illegible]	[illegible]
				Walls.	Floors.	Roof.			
LOUTH COUNTY—continued.									
[illegible]	[illegible]	[illegible]	—	—	—	—	—	—	[illegible]
MEATH COUNTY.									
[illegible]	[illegible]	[illegible]	—	—	—	—	—	—	[illegible]
[illegible]	[illegible]	[illegible]	—	—	—	—	—	—	[illegible]
[illegible]	[illegible]	[illegible]	—	—	—	—	—	—	[illegible]
[illegible]	[illegible]	[illegible]	—	—	—	—	—	—	[illegible]
[illegible]	[illegible]	—	[illegible]	[illegible]	[illegible]	[illegible]	[illegible]	[illegible]	[illegible]
			[illegible]	Do.	Do.	Do.	Do.	Do.	
			[illegible]	Do.	Do.	Do.	Do.	Do.	
			[illegible]	Do.	Do.	Do.	Do.	Do.	
			[illegible]	Do.	Do.	Do.	Do.	Do.	
			[illegible]	Do.	Do.	Do.	Do.	Do.	
			[illegible]	Do.	Do.	Do.	Do.	Do.	
			[illegible]	Do.	Do.	Do.	Do.	Do.	
[illegible]	—	[illegible]	—	—	—	—	—	—	[illegible]
[illegible]	[illegible]	[illegible]	—	—	—	—	—	—	[illegible]
[illegible]	[illegible]	[illegible]	[illegible]	[illegible]	[illegible]	[illegible]	[illegible]	[illegible]	[illegible]

LEINSTER—continued.

[illegible]	[illegible]	[illegible]
[illegible]	Same.	[illegible]
[illegible]	Same.	[illegible]
—	Same.	—
From 4 days to 6 weeks.	Old [illegible]	[illegible]
[illegible]	Old [illegible]	[illegible]
—	[illegible]	[illegible]

[illegible]	[illegible]	[illegible]	[illegible]	[illegible]	[illegible]	[illegible]	[illegible]
	[illegible]	—	—	[illegible]	—	[illegible]	[illegible]
[illegible]	[illegible]	—	—	[illegible]	—	[illegible]	[illegible]
[illegible]	[illegible]	[illegible]	[illegible]	[illegible]	[illegible]	[illegible]	[illegible]
[illegible]	[illegible]	[illegible]	—	[illegible]	—	[illegible]	[illegible]
[illegible]	[illegible]	[illegible]	—	[illegible]	—	[illegible]	[illegible]
[illegible]	[illegible]	[illegible]	[illegible]	[illegible]	[illegible]	[illegible]	[illegible]
[illegible]	[illegible]	[illegible]	[illegible]	[illegible]	[illegible]	Rarely heated.	Yes. [illegible]
[illegible]	[illegible]	[illegible]	—	[illegible]		[illegible]	Yes.
[illegible]	[illegible]	[illegible]	[illegible]	[illegible]	[illegible]	[illegible]	[illegible]
[illegible]	[illegible]	[illegible]	[illegible]	[illegible]	[illegible]	[illegible]	

LEINSTER—continued.

[illegible]	[illegible]	Tenancies		[illegible]
		[illegible]	[illegible]	
[illegible]	[illegible]	[illegible]	[illegible]	[illegible]
[illegible]	[illegible]	[illegible]	[illegible]	[illegible]
[illegible]	[illegible]	Not taken.		[illegible]
[illegible]	[illegible]	[illegible]	[illegible]	[illegible]
[illegible]	[illegible]	[illegible]	[illegible]	[illegible]
[illegible]	[illegible]	[illegible]	[illegible]	[illegible]
[illegible]	[illegible]	[illegible]	[illegible]	[illegible]
[illegible]	[illegible]	Suspended.		[illegible]

[illegible]	[illegible]	[illegible]	[illegible]	[illegible]			[illegible]	[illegible]	[illegible]
				[illegible]	[illegible]	[illegible]			
KENT COUNTY—continued.									
[illegible]	—	[illegible]	—	—	—	—	—	—	[illegible]
[illegible]	—	[illegible]	—	—	—	—	—	—	[illegible]
[illegible]	—	[illegible]	—	—	—	—	—	—	[illegible]
[illegible]	—	[illegible]	—	—	—	—	—	—	[illegible]
[illegible]	—	[illegible]	—	—	—	—	—	—	[illegible]
[illegible]	—	[illegible]	—	—	—	—	—	—	[illegible]
[illegible]	—	[illegible]	[illegible]	[illegible]	[illegible]	[illegible]	[illegible]	[illegible]	[illegible]
[illegible]	—	[illegible]	—	—	—	—	—	—	[illegible]
[illegible]	—	[illegible]	—	—	—	—	—	—	[illegible]
[illegible]	—	[illegible]	—	—	—	—	—	—	[illegible]

Road.	[illegible]	[illegible]	[illegible]
...	...	...	[illegible]
...	...	...	[illegible]
...	...	...	[illegible]
...	...	...	[illegible]
...	...	...	[illegible]
[illegible]	[illegible]	[illegible]	[illegible]
[illegible]	[illegible]	[illegible]	[illegible]
[illegible]	[illegible]	[illegible]	[illegible]
...	...	...	[illegible]

LEINSTER—continued.

[illegible]	[illegible]	Temperature.		[illegible]	[illegible]	Remarks.
		[illegible]	[illegible]			
—	—	—	—	[illegible]	[illegible]	[illegible]
[illegible]	[illegible]	—	—	[illegible]	[illegible]	[illegible]
[illegible]	[illegible]	[illegible]	[illegible]	[illegible]	[illegible]	[illegible]
[illegible]	[illegible]	—	—	—	[illegible]	[illegible]
[illegible]	[illegible]	[illegible]	[illegible]	[illegible]	[illegible]	[illegible]
[illegible]	[illegible]	[illegible]	[illegible]	[illegible]	[illegible]	[illegible]
[illegible]	[illegible]	—	—	[illegible]	[illegible]	[illegible]
—	[illegible]	—	—	—	[illegible]	[illegible]
[illegible]	[illegible]	[illegible]	[illegible]	[illegible]	[illegible]	[illegible]

LEINSTER—continued.

Number of [illegible]	[illegible]	Temperature.	
		Greatest heat.	Average heat for first [illegible] days.
[illegible]	[illegible]	Not [illegible].	
—	—	[illegible] degrees	—
—	[illegible]	[illegible]	[illegible]
[illegible]	[illegible]	Not ascertained.	

PROVINCE OF

PROVINCE OF

AGRICULTURAL STATISTICS FOR THE YEAR 188[illegible]

PROVINCE OF

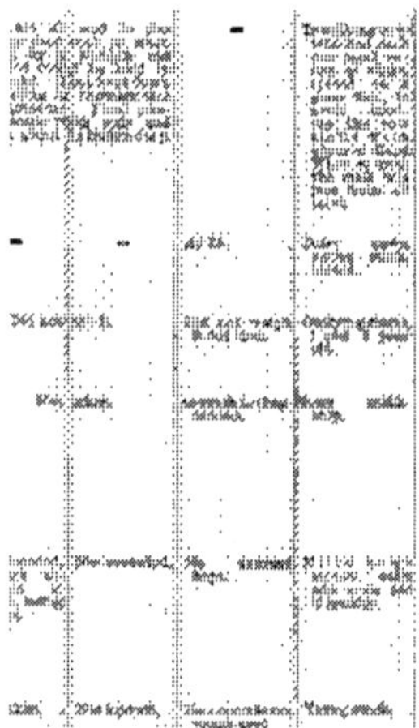

—	—	[illegible]
[illegible]	[illegible]	
—	—	[illegible]
[illegible]	[illegible]	[illegible]
—	—	[illegible]
—	—	[illegible]
—	—	[illegible]

[illegible]—continued.

[illegible]	[illegible]	Temperature [?]		[illegible]	[illegible]	Remarks
		Greatest [?]	Average [?]			
[illegible]	[illegible]	[illegible]	[illegible]	[illegible]	[illegible]	[illegible]
[illegible]	[illegible]	[illegible]		[illegible]	[illegible]	[illegible]
[illegible]	[illegible]	Not known		[illegible]	[illegible]	[illegible]
[illegible]	[illegible]	[illegible]	[illegible]	[illegible]	[illegible]	[illegible]
[illegible]	[illegible]	[illegible]	[illegible]	[illegible]	[illegible]	[illegible]
[illegible]	[illegible]	[illegible]	[illegible]	[illegible]	[illegible]	[illegible]
[illegible]	[illegible]	[illegible]	[illegible]	[illegible]	[illegible]	[illegible]
[illegible]	[illegible]	[illegible]		[illegible]	[illegible]	[illegible]
[illegible]	[illegible]	[illegible]	[illegible]	[illegible]	[illegible]	[illegible]

[illegible]—continued.

[illegible]	[illegible]	[illegible]
		[illegible]
[illegible]	[illegible]	[illegible]
[illegible]	[illegible]	[illegible]
[illegible]	[illegible]	[illegible]
[illegible]	—	[illegible]

PROVINCE OF

[illegible]	[illegible]	[illegible]	[illegible]	[illegible]			[illegible]	[illegible]	[illegible]
				Walls	Floor	Roof			
TIPPERARY COUNTY—continued.									
[illegible]	[illegible]	[illegible]	[illegible]	[illegible]	[illegible]	[illegible]	[illegible]	[illegible]	[illegible]
[illegible]	[illegible]	[illegible]	[illegible]	[illegible]	[illegible]	[illegible]	[illegible]	[illegible]	[illegible]
[illegible]	[illegible]	[illegible]	[illegible]	[illegible]	[illegible]	[illegible]	[illegible]	[illegible]	[illegible]
[illegible]	[illegible]	[illegible]	[illegible]	[illegible]	[illegible]	[illegible]	[illegible]	[illegible]	[illegible]
[illegible]	[illegible]	[illegible]	[illegible]	[illegible]	[illegible]	[illegible]	[illegible]	[illegible]	[illegible]
[illegible]	[illegible]	[illegible]	[illegible]	[illegible]	[illegible]	[illegible]	[illegible]	[illegible]	[illegible]
[illegible]	[illegible]	[illegible]	[illegible]	[illegible]	[illegible]	[illegible]	[illegible]	[illegible]	[illegible]
[illegible]	[illegible]	[illegible]	[illegible]	[illegible]	[illegible]	[illegible]	[illegible]	[illegible]	[illegible]
WATERFORD COUNTY.									
[illegible]	[illegible]	[illegible]	[illegible]	[illegible]	[illegible]	[illegible]	[illegible]	[illegible]	[illegible]
[illegible]	[illegible]	[illegible]	[illegible]	[illegible]	[illegible]	[illegible]	[illegible]	[illegible]	[illegible]

[illegible]	[illegible]	[illegible]	[illegible]	[illegible]	[illegible]	[illegible]	[illegible]
[illegible]	—	—	—	—	—	—	[illegible]
[illegible]	[illegible]	[illegible]	[illegible]	[illegible]	Do.	[illegible]	[illegible]

PROVINCE O[illegible]

[illegible]	[illegible]	[illegible]	[illegible]	[illegible]	[illegible]	[illegible]	[illegible]
	[illegible]	—	[illegible]	—	—	Do.	
[illegible]	[illegible]	[illegible]	[illegible]	[illegible]	[illegible]	[illegible]	Do.
[illegible]	[illegible]	[illegible]	[illegible]	[illegible]	[illegible]	[illegible]	[illegible]
[illegible]	—	—	—	—	—	[illegible]	[illegible]
[illegible]	[illegible]	[illegible]	[illegible]	[illegible]	[illegible]	[illegible]	—
[illegible]	[illegible]	—	—	—	—	—	Do.
[illegible]	[illegible]	[illegible]	[illegible]	[illegible]	[illegible]	[illegible]	Do.

[illegible]	[illegible]	[illegible]	[illegible]	[illegible]	[illegible]
[illegible]	[illegible]	[illegible]	[illegible]	[illegible]	[illegible]
[illegible]	[illegible]	[illegible]	[illegible]	[illegible]	[illegible]
—	—	—	—	—	—
[illegible]	[illegible]	[illegible]	[illegible]	[illegible]	[illegible]
[illegible]	[illegible]	[illegible]	[illegible]	[illegible]	[illegible]
[illegible]		—	—	—	—
—	—		—	—	—
—	—	—	—	—	—

	—	—	—			[illegible]
[illegible]	[illegible]	[illegible]	[illegible]	[illegible]	[illegible]	...
[illegible]	[illegible]	[illegible]	[illegible]	[illegible]	[illegible]	[illegible]
—	—	—	—	—	[illegible]	[illegible]
[illegible]	[illegible]	[illegible]	[illegible]	[illegible]	[illegible]	[illegible]
—	—	—	—	—	—	[illegible]

[illegible]	[illegible]			[illegible]	[illegible]	[illegible]
	Walls.	Floor.	Roof.			
[illegible]	[illegible]	[illegible]	[illegible]	[illegible]	[illegible]	[illegible]
[illegible]	Stone. Do.	[illegible]	[illegible]	[illegible]	[illegible]	[illegible]
—	—	—	—	—	—	[illegible]
[illegible]	Stone and lime.	Concrete.	[illegible]	[illegible]	[illegible]	[illegible]
[illegible]	Do.	[illegible]	[illegible]	[illegible]	[illegible]	
—	—	—	—	—	—	[illegible]
—		—		—	—	[illegible]
—	—		—	—	—	[illegible]

[illegible]	[illegible]	[illegible]	[illegible]	[illegible]	[illegible]	[illegible]	[illegible]
[illegible]	[illegible]	[illegible]	[illegible]	[illegible]	[illegible]	[illegible]	[illegible]
[illegible]	[illegible]	[illegible]	[illegible]	[illegible]	[illegible]	[illegible]	[illegible]
[illegible]	[illegible]	[illegible]	[illegible]	[illegible]	[illegible]	[illegible]	[illegible]
[illegible]	—	—	—	—	—	—	[illegible]
[illegible]	[illegible]	[illegible]	[illegible]	[illegible]	[illegible]	[illegible]	[illegible]
[illegible]	—	—	—	—	[illegible]	[illegible]	[illegible]
[illegible]	—	—	—	—	—	—	[illegible]

[illegible]—continued.

[illegible]	[illegible]	[illegible]
[illegible]	[illegible]	[illegible]
[illegible]	[illegible]	[illegible]
[illegible]	[illegible]	[illegible]

Dimensions of Silos—Length, Breadth, Depth.	Materials of Silo.			Whether Drained or not.
	Walls.	Floor.	Roof.	
—	—	—	—	—
[illegible] feet by [illegible] feet by [illegible] feet deep.	[illegible], with cement [illegible].	Concrete [illegible].	Galvanised iron.	Yes.
—	—	—	—	—
[illegible] feet by [illegible] feet, [illegible] feet deep.	Stone with cement lining.	Concrete.	Wood.	No.
—	—	—	—	—
—	—	—	—	—
[illegible] feet by [illegible] feet, [illegible] feet deep.	Stone cemented [illegible]	Earth.	[illegible].	[illegible].
—	—	—	—	—
—	—	—	—	—
[illegible] feet by [illegible] feet, [illegible] feet deep.	Brick and cement.	Concrete.	[illegible]	[illegible]

PROVINCE OF

Dimensions of Silo.—Length, Breadth, Depth.	Materials of Silo.			Whether Covered or not.	[illegible]	[illegible]
	Walls.	Floor.	Roof.			

CHHNAUGHT—continued.

[illegible]	[illegible]	[illegible]
		[illegible]
[illegible]	[illegible]	[illegible]

CONNAUGHT—*continued.*

[illegible]	[illegible]	[illegible]		[illegible]	[illegible]	Remarks.
		[illegible]	[illegible]			
[illegible]	[illegible]	[illegible]	[illegible]	[illegible]	[illegible]	[illegible]
[illegible]	[illegible]	[illegible]		[illegible]	[illegible]	[illegible]
[illegible]	[illegible]	—	—	[illegible]	[illegible]	[illegible]
[illegible]	[illegible]	[illegible]	[illegible]	[illegible]	[illegible]	[illegible]
[illegible]	[illegible]	[illegible]		—	[illegible]	[illegible]
[illegible]	[illegible]	[illegible]	[illegible]	[illegible]	[illegible]	[illegible]
[illegible]	[illegible]	[illegible]	[illegible]	[illegible]	[illegible]	[illegible]
[illegible]	[illegible]	[illegible]		[illegible]	[illegible]	[illegible]

				[illegible]
–	–	–	–	[illegible]
[illegible]	[illegible]	Yes.	Early before	[illegible]
Do.	Do.	Do.	Do.	
Do.	Do.	Do.	Do.	
Do.	Do.	Do.	Do.	
[illegible]	[illegible]	[illegible]	[illegible]	–
[illegible]	Do.	Do.	Do.	
[illegible]	[illegible]	Do.	Do.	
[illegible]	[illegible]	Do.	[illegible]	
–	–	–	–	[illegible]
–	–	–	–	[illegible]
[illegible]	[illegible]	[illegible]	[illegible]	[illegible]
Do.	Do.	Do.	Do.	

CONNAUGHT—*continued*

[illegible]	[illegible]	Temperature.		[illegible]	[illegible]	[illegible]
		[illegible]	[illegible]			
[illegible]	[illegible]	[illegible]		[illegible]	[illegible]	[illegible]

THE WEATHER.

Abstract of Meteorological Observations registered at the Ordnance Survey Office (Height above the Sea 162½ Feet) Phœnix Park, Dublin, during the year 1869:—

The barometer stood highest in 1869, on the 5th December, at 9 P.M., the air being calm, when it was 30·718 inches; it was lowest at 9 P.M. on 19th March, when it was 28·948 inches. The highest temperature of the air during the year was 78·2 degrees of Fahrenheit on 22nd June, and the lowest 17·2 degrees on 11th February. The greatest quantity of rain which fell in a day (24 hours) was 1·319 inches on 18th August, with wind E.N.E. The point from which the wind chiefly prevailed was the W.; it blew from that direction on 136 days, at 9 A.M. The strongest wind was from the W. on the 7th October, and from the S.W. on the 19th December, when the pressure on each day was 12·5 lbs. per square foot.

1869.	Barometer. Corrected for Temperature and Reduced to 32° Fahr. Mean. 9 A.M.	9 P.M.	Mean.	Range.	Highest in Month.	Lowest in Month.	Range.	Thermometer. Self-Registering Thermometers. Highest in Month.	Lowest in Month.	Range.	Mean. Of all Maxima.	Of all Minima.	Daily.	Range.	Hygrometer. 9 A.M. Mean. Dry Bulb.	Wet Bulb.	Dew Point.	Elastic Force of Vapour.	Humidity.
January,	[illegible]	[illegible]	[illegible]	[illegible]	[illegible]	[illegible]	[illegible]	[illegible]	[illegible]	[illegible]	[illegible]	[illegible]	[illegible]	[illegible]	[illegible]	[illegible]	[illegible]	[illegible]	[illegible]
February,	[illegible]	[illegible]	[illegible]	[illegible]	[illegible]	[illegible]	[illegible]	[illegible]	[illegible]	[illegible]	[illegible]	[illegible]	[illegible]	[illegible]	[illegible]	[illegible]	[illegible]	[illegible]	[illegible]
March,	[illegible]	[illegible]	[illegible]	[illegible]	[illegible]	[illegible]	[illegible]	[illegible]	[illegible]	[illegible]	[illegible]	[illegible]	[illegible]	[illegible]	[illegible]	[illegible]	[illegible]	[illegible]	[illegible]
April,	[illegible]	[illegible]	[illegible]	[illegible]	[illegible]	[illegible]	[illegible]	[illegible]	[illegible]	[illegible]	[illegible]	[illegible]	[illegible]	[illegible]	[illegible]	[illegible]	[illegible]	[illegible]	[illegible]
May,	[illegible]	[illegible]	[illegible]	[illegible]	[illegible]	[illegible]	[illegible]	[illegible]	[illegible]	[illegible]	[illegible]	[illegible]	[illegible]	[illegible]	[illegible]	[illegible]	[illegible]	[illegible]	[illegible]
June,	[illegible]	[illegible]	[illegible]	[illegible]	[illegible]	[illegible]	[illegible]	[illegible]	[illegible]	[illegible]	[illegible]	[illegible]	[illegible]	[illegible]	[illegible]	[illegible]	[illegible]	[illegible]	[illegible]
July,	[illegible]	[illegible]	[illegible]	[illegible]	[illegible]	[illegible]	[illegible]	[illegible]	[illegible]	[illegible]	[illegible]	[illegible]	[illegible]	[illegible]	[illegible]	[illegible]	[illegible]	[illegible]	[illegible]
August,	[illegible]	[illegible]	[illegible]	[illegible]	[illegible]	[illegible]	[illegible]	[illegible]	[illegible]	[illegible]	[illegible]	[illegible]	[illegible]	[illegible]	[illegible]	[illegible]	[illegible]	[illegible]	[illegible]
September,	[illegible]	[illegible]	[illegible]	[illegible]	[illegible]	[illegible]	[illegible]	[illegible]	[illegible]	[illegible]	[illegible]	[illegible]	[illegible]	[illegible]	[illegible]	[illegible]	[illegible]	[illegible]	[illegible]
October,	[illegible]	[illegible]	[illegible]	[illegible]	[illegible]	[illegible]	[illegible]	[illegible]	[illegible]	[illegible]	[illegible]	[illegible]	[illegible]	[illegible]	[illegible]	[illegible]	[illegible]	[illegible]	[illegible]
November,	[illegible]	[illegible]	[illegible]	[illegible]	[illegible]	[illegible]	[illegible]	[illegible]	[illegible]	[illegible]	[illegible]	[illegible]	[illegible]	[illegible]	[illegible]	[illegible]	[illegible]	[illegible]	[illegible]
December,	[illegible]	[illegible]	[illegible]	[illegible]	[illegible]	[illegible]	[illegible]	[illegible]	[illegible]	[illegible]	[illegible]	[illegible]	[illegible]	[illegible]	[illegible]	[illegible]	[illegible]	[illegible]	[illegible]
Total,	[illegible]	[illegible]	[illegible]	–	–	–	–	–	–	–	[illegible]	[illegible]	[illegible]	–	[illegible]	[illegible]	[illegible]	[illegible]	[illegible]
Mean,	[illegible]	[illegible]	[illegible]	–	–	–	–	–	–	–	[illegible]	[illegible]	[illegible]	–	[illegible]	[illegible]	[illegible]	[illegible]	[illegible]

1869.	Rain. Number of days on which rain fell.	Total amount in inches.	Greatest amount in 24 hours.	Ozone.	Clouds. Average amount.	Calm days.	Wind. Number of days it blew from each direction at 9 A.M. and the mean pressure in lbs. per square foot. N.	Pressure.	N.E.	Pressure.	E.	Pressure.	S.E.	Pressure.	S.	Pressure.	S.W.	Pressure.	W.	Pressure.	N.W.	Pressure.
January,	[illegible]	[illegible]	[illegible]	[illegible]	[illegible]	[illegible]	[illegible]	[illegible]	[illegible]	[illegible]	[illegible]	[illegible]	[illegible]	[illegible]	[illegible]	[illegible]	[illegible]	[illegible]	[illegible]	[illegible]	[illegible]	[illegible]
February,	[illegible]	[illegible]	[illegible]	[illegible]	[illegible]	[illegible]	[illegible]	[illegible]	[illegible]	[illegible]	[illegible]	[illegible]	[illegible]	[illegible]	[illegible]	[illegible]	[illegible]	[illegible]	[illegible]	[illegible]	[illegible]	[illegible]
March,	[illegible]	[illegible]	[illegible]	[illegible]	[illegible]	[illegible]	[illegible]	[illegible]	[illegible]	[illegible]	[illegible]	[illegible]	[illegible]	[illegible]	[illegible]	[illegible]	[illegible]	[illegible]	[illegible]	[illegible]	[illegible]	[illegible]
April,	[illegible]	[illegible]	[illegible]	[illegible]	[illegible]	[illegible]	[illegible]	[illegible]	[illegible]	[illegible]	[illegible]	[illegible]	[illegible]	[illegible]	[illegible]	[illegible]	[illegible]	[illegible]	[illegible]	[illegible]	[illegible]	[illegible]
May,	[illegible]	[illegible]	[illegible]	[illegible]	[illegible]	[illegible]	[illegible]	[illegible]	[illegible]	[illegible]	[illegible]	[illegible]	[illegible]	[illegible]	[illegible]	[illegible]	[illegible]	[illegible]	[illegible]	[illegible]	[illegible]	[illegible]
June,	[illegible]	[illegible]	[illegible]	[illegible]	[illegible]	[illegible]	[illegible]	[illegible]	[illegible]	[illegible]	[illegible]	[illegible]	[illegible]	[illegible]	[illegible]	[illegible]	[illegible]	[illegible]	[illegible]	[illegible]	[illegible]	[illegible]
July,	[illegible]	[illegible]	[illegible]	[illegible]	[illegible]	[illegible]	[illegible]	[illegible]	[illegible]	[illegible]	[illegible]	[illegible]	[illegible]	[illegible]	[illegible]	[illegible]	[illegible]	[illegible]	[illegible]	[illegible]	[illegible]	[illegible]
August,	[illegible]	[illegible]	[illegible]	[illegible]	[illegible]	[illegible]	[illegible]	[illegible]	[illegible]	[illegible]	[illegible]	[illegible]	[illegible]	[illegible]	[illegible]	[illegible]	[illegible]	[illegible]	[illegible]	[illegible]	[illegible]	[illegible]
September,	[illegible]	[illegible]	[illegible]	[illegible]	[illegible]	[illegible]	[illegible]	[illegible]	[illegible]	[illegible]	[illegible]	[illegible]	[illegible]	[illegible]	[illegible]	[illegible]	[illegible]	[illegible]	[illegible]	[illegible]	[illegible]	[illegible]
October,	[illegible]	[illegible]	[illegible]	[illegible]	[illegible]	[illegible]	[illegible]	[illegible]	[illegible]	[illegible]	[illegible]	[illegible]	[illegible]	[illegible]	[illegible]	[illegible]	[illegible]	[illegible]	[illegible]	[illegible]	[illegible]	[illegible]
November,	[illegible]	[illegible]	[illegible]	[illegible]	[illegible]	[illegible]	[illegible]	[illegible]	[illegible]	[illegible]	[illegible]	[illegible]	[illegible]	[illegible]	[illegible]	[illegible]	[illegible]	[illegible]	[illegible]	[illegible]	[illegible]	[illegible]
December,	[illegible]	[illegible]	[illegible]	[illegible]	[illegible]	[illegible]	[illegible]	[illegible]	[illegible]	[illegible]	[illegible]	[illegible]	[illegible]	[illegible]	[illegible]	[illegible]	[illegible]	[illegible]	[illegible]	[illegible]	[illegible]	[illegible]
Total,	[illegible]	[illegible]	–	[illegible]	[illegible]	[illegible]	[illegible]	[illegible]	[illegible]	[illegible]	[illegible]	[illegible]	[illegible]	[illegible]	[illegible]	[illegible]	[illegible]	[illegible]	[illegible]	[illegible]	[illegible]	[illegible]
Mean,	–	–	–	[illegible]	[illegible]	[illegible]	–	[illegible]	–	[illegible]	–	[illegible]	–	[illegible]	–	[illegible]	–	[illegible]	–	[illegible]	–	[illegible]

METEOROLOGICAL OBSERVATIONS
FOR EACH MONTH OF THE YEAR 1889.

By J. W. MOORE, Esq., M.D., [illegible], F.R., [illegible], etc.

(Extracted from the *Dublin Journal of Medical Science*.)

JANUARY.—As happened also in January, 1888, during the greater part of the month, the weather was [illegible] changeable in Ireland and in Scotland, colder in England, and very cold in France and Germany. The type of distribution of atmospherical pressure was chiefly anticyclonic in the south and south-east, cyclonic in the far north and north-west.

In Dublin the mean temperature ([illegible]°) was perceptibly above the average ([illegible]°); the mean dry bulb readings at 9 a.m. and 9 p.m. were [illegible]°. In the twenty-four years ending with 1888, January was coldest in 1881 (M. T. = [illegible]°) and warmest in 1873 (M. T. = [illegible]°). In 1867, the M. T. was [illegible]°, and in 1869 it was [illegible]°. In 1871 and in 1880, the M. T. was [illegible]; in the year 1879 (the cold year), it was [illegible]°. In 1888, the M. T. was [illegible]°. As a general rule, January in Dublin is not colder, but rather a shade warmer, than December. This is owing to the full development in January of a winter area of low pressure over the Atlantic, to the north-westward of the British Isles, and to a resulting prevalence of S.W. winds in their vicinity. January, 1888, proved an exception to this rule, the M. T. being [illegible]° below that of December, 1887 ([illegible]°).

The mean height of the barometer was 30·[illegible] inches, or as much as 0·[illegible] inch above the average value for January—namely, 29·[illegible] inches. The mercury rose to 30·[illegible] inches at 9 a.m. of the 3rd, and fell to 29·[illegible] inches at 8 p.m. of the [illegible]th. The observed range of atmospherical pressure was, therefore, as much as 1·[illegible] inches—that is, a little over one inch and one-half. The mean temperature deduced from daily readings of the dry bulb thermometer at 9 a.m. and 9 p.m. was [illegible]°, or [illegible]° below the value for December, 1888; that calculated by Kaemtz's formula—viz., min. + (max.—min. × ·41) = Mean Temp.—from the means of the daily maxima and minima was [illegible]°, or [illegible]° above the average mean temperature for January, calculated in the same way, in the twenty years, 1865-84, inclusive ([illegible]°). The arithmetical mean of the maximal and minimal readings was [illegible]°, compared with a 23 years' average of [illegible]°. On the 15th, the thermometer in the screen rose to [illegible]°—wind W.S.W.; on the 3rd, the temperature fell to [illegible]°—wind also W.S.W. The minimum on the grass was [illegible]° on the same date. The rainfall was [illegible] inches, distributed over 15 days. The average rainfall for January in the twenty-three years, 1866-87, inclusive, was 2·[illegible] inches, and the average number of rainy days was 17·[illegible]. The rainfall and the rainy days, therefore, were both slightly below the average. In 1877 the rainfall in January was very large—[illegible] inches in 25 days; in 1883 also [illegible] inches fell—on, however, only [illegible] days. On the other hand, in 1876, only ·[illegible] of an inch was measured on but [illegible] days; and in 1880, the rainfall was only ·[illegible] of an inch on but [illegible] days. In January, 1884, [illegible] inches of rain were measured on as many as 26 days, in 1887 (the dry year) [illegible] inches fell on 16 days, and in 1888 [illegible] inches fell on 9 days.

A lunar halo was seen on the 14th. The atmosphere was foggy on each of the first six days, as also on the [illegible]. High winds were noted on 8 days, reaching the force of a gale on two days—the 11th and 15th. Hail fell on the 9th and 11th, and snow or sleet also on the 11th. Temperature exceeded 50° in the screen on 8 days, compared with 10 days in December, 1888; while it fell to or below 32° in the screen on only 3 nights compared with 5 nights in December. The minima on the grass were 32°, or less, on 15 nights compared with [illegible] nights in December.

Cold, quiet, foggy weather prevailed during the greater part of the week ended January 5th, 1889, in the centre and east of Ireland, as well as in England. In the west of Ireland and in Scotland S.W. winds and cloudy, mild weather were experienced. From Tuesday, the 1st, to the close of the week a band of high atmospherical pressure extended from North Germany across England and Ireland, while the barometer was low over the Norwegian Sea and Scandinavia in the north and over the Mediterranean in the south. Sharp frosts prevailed within the area of high pressure, but it was very mild on the Atlantic coasts of Ireland, Scotland, and Norway. In Dublin the barometer rose to 30·[illegible] inches at 9 a.m. of Thursday, the 3rd (wind W.S.W.). Thick fogs formed on Friday and Saturday. Fogs were observed on Wednesday and Saturday. Rain fell on Saturday night to the amount of ·021 inch.

As regards the second week (6th—12th inclusive), except on Thursday, the 10th, which was a fine, dry, bracing day, the weather was either cloudy and damp, or rainy. At the beginning of the week intense frost prevailed in central and eastern England—at 8 a.m. of Sunday, the 6th, the thermometer read 15° at Cambridge, 21° at Loughborough, and [illegible]° in London. This was a dull, damp, foggy day in Dublin. Conditions for southerly winds became pronounced over Western Europe, temperature rose and rain fell very generally. On Wednesday the wind shifted to W. and N.W. in Ireland with a downpour of rain and hail and an abrupt fall of temperature. On Friday afternoon there was a S.E. gale and immense quantities of rain, hail, and sleet fell in Dublin and its vicinity. Saturday was a dull day, with snow in England. In Dublin the mean height of the barometer was 29·[illegible] inches, pressure varying from 30·[illegible] inches at 9 a.m. of Sunday (wind, W.) to 29·[illegible] inches at 3 p.m. of Wednesday (wind, W.N.W.). The mean dry bulb temperature at 9 a.m. and 9 p.m. was [illegible]°, or [illegible]° above that for the previous week. The mean of the daily highest and lowest temperatures was [illegible]°. The absolute maximum in the screen was [illegible]° on Tuesday (wind, S.); the absolute minimum was [illegible]° on Friday (wind calm). Rain fell on five days (including hail on two days and sleet on one day) to the large amount of 1·[illegible] inches. Of this quantity, 1·[illegible] inches were measured on Friday and ·[illegible] inch on Tuesday. There was a fog on Sunday, and a gale was recorded on Friday.

In the third week (13th–19th inclusive) open, but changeable weather was prevalent. Atmospherical pressure was generally lower to the northwestward than to the southeastward, and in consequence the winds were chiefly southerly to westerly. On Friday morning the force of a gale was reached, but otherwise the winds were not strong. On Wednesday a V-shaped depression crossed Ireland, causing a sudden shift of wind from S. to N.W.; but it soon backed again to S.W. A partial eclipse of the moon was seen in a clear sky on the morning of Thursday, the 17th. Showers fell on the afternoons of Friday and Saturday. Very little frost occurred in the British Isles, and on Friday very high temperatures were recorded in Ireland and Scotland. The weather remained persistently cold in Germany and eastern France. At 8 a.m. of Tuesday the thermometer was 4° at Berlin. The barometer was extremely high over Russia during this week; at Moscow it rose to 31·23 inches on Monday morning. In Dublin the mean atmospherical pressure was 30·124 inches; the extremes were—lowest, 29·780 inches, at 9 a.m. of Wednesday (wind, S.S.E.); highest, 30·401 inches, at 9 a.m. of Saturday (wind, W. by S.). The mean dry bulb temperature at 9 a.m. and 9 p.m. was 43·6°. The arithmetical mean of the highest and lowest daily temperatures was 44·6°. On Friday the screened thermometers rose to 56·8° (wind, W.S.W.); on Thursday, they fell to 35·5° (wind, S.W.). Rain fell on four days to the amount of ·179 inch—the maximal fall in 24 hours was ·080 inch on Tuesday (wind, S.). A lunar halo was seen on Monday evening, the 14th.

Throughout the week ending Saturday, the 26th, an anticyclone, or area of high atmospherical pressure, lay over Ireland and the Atlantic Ocean off the west coast of this country. At the same time, extensive and deep depressions passed eastwards across Scandinavia and Lapland, while areas of relatively low pressure lay also over the Mediterranean Basin. Light or moderate N.W. (N. to W.) winds prevailed in the United Kingdom, accompanied by comparatively mild and dry, but cloudy weather. At 8 a.m. of Friday the barometer varied from 30·57 inches at Valentia Island, in Kerry, to 29·59 inches at Haparanda, on the Gulf of Bothnia—rather steep gradients for N.W. winds consequently prevailed over Northwestern Europe. In Dublin the mean height of the barometer during the week was no less than 30·510 inches, pressure ranging from 30·347 inches at 9 a.m. of Sunday (wind, W.), to 30·680 inches at 9 a.m. of Tuesday (wind, W.N.W.). The mean dry bulb temperature at 9 a.m. and 9 p.m. was 44·0°; the mean of the highest and lowest daily temperatures was 44·7°. The thermometers in the screen rose to 50·1° on Saturday (wind, W.), having fallen to 35·0° on Tuesday (wind, W.N.W.). Rain fell in measurable quantity on two days—Tuesday and Saturday—the total precipitation being ·048 inch, of which ·038 inch fell on Saturday evening, as a shallow secondary depression passed over Dublin from N.W. to S.E. Very cloudy skies prevailed during the week, the percentage amount of cloud at 9 a.m. and 9 p.m. being 87. Only on Tuesday was there much bright sunshine. The mildness of the northerly and northwesterly winds which prevailed was due to their origin in an oceanic anticyclone.

The last five days of the month were chiefly mild, changeable, and cloudy, and rain fell frequently although not in large quantity. On Tuesday, the 29th, a remarkable series of V-shaped depressions crossed the United Kingdom, causing a shift of wind from S.W. to N.W., squalls, showers, and a brisk fall of temperature. At the close of the month steep gradients for westerly winds prevailed over Northwestern Europe, and the weather was generally mild and damp.

At Greystones, Co. Wicklow, the rainfall in January, 1889, was 3·67 inches, distributed over only 8 days. Of this quantity 1·64 inches fell on the 11th, and ·98 of an inch on the 5th.

FEBRUARY.—This was an unsettled, windy, wet, and cold month. North-westerly winds preponderated, and while the barometer was often high off the W. and S.W. of Ireland, deep atmospherical depressions passed south-eastwards across Scandinavia and the North Sea, producing cold N.W. winds and frequent showers of sleet and hail. On the 10th a severe snowstorm occurred, and the weather was generally inclement.

In Dublin the mean temperature (40·3°) was much below the average (43·0°); the mean dry bulb readings at 9 a.m. and 9 p.m. were 39·6°. In the twenty-four years ending with 1888, February was coldest in 1873 (M. T. = 37·9°) and warmest in 1869 (M. T. = 46·7°). In 1886, the M. T. was 39·7°, in the year 1879 (the cold year), it was 40·1°, and in 1888 it was as low as 38·6°. As a general rule, February in Dublin is only a shade colder than March.

The mean height of the barometer was 29·891 inches, or 0·129 inch above the average value for February—namely, 29·862 inches. The mercury rose to 30·474 inches at 9 p.m. of the 4th, and fell to 29·189 inches at 6 p.m. of the 10th. The observed range of atmospherical pressure was, therefore, 1·285 inches—that is, a little over one inch and a quarter. The mean temperature deduced from daily readings of the dry bulb thermometer at 9 a.m. and 9 p.m. was 39·6°, or 2·3° below the value for January, 1889; that calculated by Kaemtz's formula from the means of the daily maxima and minima was 39·4°, or 3·1° below the average mean temperature for February, calculated in the same way, in the twenty years, 1865–84, inclusive (42·5°). The arithmetical mean of the maximal and minimal readings was 40·3°, compared with a 23 years' average of 43·0°. On the 1st, the thermometer in the screen rose to 53·0°—wind W.S.W.; on the 11th, the temperature fell to 31·7°—wind W.N.W. The minimum on the grass was 14·8° also on the latter date. The rainfall was 2·449 inches, distributed over no fewer than 20 days. The average rainfall for February in the twenty-three years, 1865–87, inclusive, was 2·183 inches, and the average number of rainy days was 17·3. The rainfall and the rainy days, therefore, were both considerably above the average. In 1883 the rainfall in February was large—3·753 inches on 17 days; in 1879 also 3·706 inches fell on 23 days. On the other hand, in 1873, only ·925 of an inch was measured on but 6 days; and in 1887, only ·541 inch of rain fell on 11 days. The rainfall in 1887 was much the smallest recorded in February for very many years. Snow or sleet fell on the 2nd, 8th, 9th, 10th, 25th, 26th, 27th, and 28th. Hail fell on the 1st, 4th, 8th, 10th, 24th, 26th, and 27th. The atmosphere was foggy on the 15th, 16th, 27th, and 28th. High winds were noted on 13 days, reaching the force of a gale on five days

—the 1st, 2nd, 3rd, [illegible], and [illegible]. The temperature exceeded [illegible]° in the screen on [illegible] days, compared with the same number of days in January; while it fell to or below 32° in the screen on only [illegible] days, compared with only three in January, but with [illegible] in February, [illegible]. The minima on the grass were 32°, or less, on [illegible] nights, compared with [illegible] nights in January. On 7 days the thermometer did not rise to 40° in the screen.

On Friday, February 1, a fresh westerly gale prevailed, and temperature—which had been very high during the previous night ([illegible]° in Dublin)—gave way rapidly, with squalls of rain and hail. Saturday, the 2nd, proved the most inclement day as yet felt in the present winter—a fresh westerly gale blowing with frequent falls of snow and bitter cold. The thermometer rose to [illegible]° on the early morning of Friday (wind, W.S.W.), and fell to [illegible]° on Saturday morning (wind, W.). At [illegible] a.m. of this day the barometer was down to [illegible] inches at Sumburgh Head in the Shetlands. Strong N.W. gales, snow and hail, thunder and lightning prevailed in many parts of the Kingdom, and in Dublin [illegible] of an inch of rain fell, in the form of snow.

The weather was very unsettled and changeable during the week ended Saturday, the 9th. Severe, extensive, and deep depressions travelled southeastwards across Northwestern Europe, causing rapid changes of temperature, and westerly to northerly gales, with frequent showers of rain or hail and snow. On Sunday a very deep depression was passing down the North Sea, accompanied by severe northerly gales and thunder and lightning in many parts of the United Kingdom. In the rear of the disturbance the barometer rose with great rapidity—nearly an inch and a half at Aberdeen in the [illegible] hours ending 8 a.m. of Monday. During the next two days areas of low pressure crossed Scandinavia from N.W. to S.E., and the weather was dull, damp, and showery. On Wednesday night a subsidiary depression passed rapidly across the British Islands in the same direction—rain and hail squalls were reported from many stations. During Friday and Saturday another very deep depression travelled southeastwards down the North Sea, the accompanying gales being of great intensity. In Scotland snow fell heavily, with keen frost, and in Ireland the air became very dry and cold. Sharp frost set in on Saturday. In Dublin the mean height of the barometer was [illegible] inches—pressure ranging from [illegible] inches at [illegible] a.m. of Sunday (wind, W.N.W.) to [illegible] inches at [illegible] p.m. of Monday (wind, N.). The mean dry bulb temperature at 9 a.m. and 9 p.m. was [illegible]°, or 6° below the value for the previous week. The arithmetical mean of the highest and lowest daily temperatures was [illegible]°. Temperature in the screen rose to [illegible]° on Wednesday (wind, W.) and Thursday (wind, W.N.W.), and fell to [illegible]° on Saturday (wind, N.W.). Rain fell on six days to the amount of [illegible] inch, including ·117 inch referred to Wednesday. Hail and snow fell on two days.

Very changeable, inclement weather held during the greater part of the week ended Saturday, the 16th, which began with a snowstorm and ended with warmth and squalls. On Sunday two depressions crossed Ireland and England—the first from W. to E., the second from W.N.W. to E.S.E.—both travelling with great speed. In Dublin, fine dry snow began to fall at [illegible] a.m., with a temperature of [illegible]°. As the forenoon wore on, the snowfall increased. In the afternoon a thaw occurred and snow, rain and hail were followed by rain and a shift of wind from S.E. to W. In the evening a further shift of wind to N.N.E. took place and snow and hail again fell heavily. On Monday morning there was an intense frost. Tuesday was again fair and frosty, but at night a lower [illegible] [illegible] a change. Wednesday was dull and rainy, and at night there was a fresh gale from W.S.W., temperature rising to [illegible]°. Thursday and Friday were mild, quiet days, but on Friday evening there was again a lower [illegible], and Saturday was rainy and squally, [illegible]. At Dungeness, in Kent, temperature rose [illegible]° (from [illegible]° to [illegible]°) in the twenty-four hours ending 8 a.m. of Thursday. In Dublin the mean height of the barometer during the week was [illegible] inches—pressure ranging from [illegible] inches at [illegible] p.m. of Sunday (wind, W.) to [illegible] inches at [illegible] a.m. of Tuesday (wind, W.N.W.). The mean dry bulb temperature at 9 a.m. and 9 p.m. was [illegible]°, while the mean temperature derived from the daily maxima and minima was [illegible]°. The thermometer in the screen fell to [illegible]° on Monday (wind, W.N.W.) and rose to [illegible]° on Saturday (wind, W.S.W.). The rainfall amounted to [illegible] inches on [illegible] days. Of this quantity, ·742 inch was the result of melted snow and hail on Sunday.

As regards the third week (17th–23rd, inclusive), at first mild, soft, and cloudy, the weather afterwards became cooler, drier, and brighter as [illegible] gradients for northerly winds developed over Western Europe. Until Wednesday an anticyclone was found over Spain, the Bay of Biscay, and France, while areas of low pressure passed in a direction towards E. or S.E. across the Norwegian Sea and Scandinavia. Westerly winds and mild, cloudy weather prevailed during this period, except in the north of Sweden where the cold was very intense. On Thursday, the isobars ran nearly due N. and S., and the wind drew into N. with a general fall of temperature and [illegible] [illegible] in Great Britain. Next day a shallow depression passed southwards across Ireland, causing a higher temperature for the time being and showers. On Saturday the wind shifted to E. and became dry and searching—the relative humidity in Dublin falling to [illegible] per cent. at [illegible] p.m. In this city the mean height of the barometer was [illegible] inches—pressure ranging from [illegible] inches at [illegible] a.m. of Sunday (wind, W.S.W.) to [illegible] inches at [illegible] p.m. of Tuesday (wind, W.N.W.). The mean dry bulb temperature at 9 a.m. and 9 p.m. was [illegible]°, while the mean of the daily maxima and minima was [illegible]°. Temperature in the screen rose to [illegible]° on Monday (wind, S.W.) and fell to [illegible]° on Friday (wind, W.). Rain was registered on four days, the total quantity being [illegible] inch, of which [illegible] inch was credited to Friday. At 8 a.m. of Saturday, February 23rd, temperature in Western Europe ranged from [illegible]° Fahr. at Haparanda in Sweden, [illegible]° at Munich, and [illegible]° at Lyons, to [illegible]° at Belmullet and Valentia Island, Kerry.

Very inclement weather prevailed throughout the closing period of the month (from the 24th to the [illegible], inclusive). Barometrical gradients for easterly (N., to E. and S.E.) winds held over Western

Europe, so that a polar air-current held, temperature ruled very low, and snow and hail fell frequently and on the east coasts of Great Britain and Ireland in considerable quantity. In the West of Ireland fine, bright, frosty weather prevailed for the most part, but in the East clouded skies and snow were experienced. At 8 a.m. of Thursday, the 28th, the thermometer was only 21° in the screen at Parsonstown, King's Co.

The rainfall in Dublin during the two months ending February 28th has amounted to 4·682 inches on 36 days, compared with 2·346 inches on 23 days during the same period in 1888, and a 23 years' average of 4·424 inches on 35·0 days. Double as much rain as fell in January and February of last year has fallen in the same months this year.

At Greystones, Co. Wicklow, the rainfall in February, 1889, was 2·60 inches, distributed over 14 days. Of this quantity 1·12 inches fell in the form of snow on the 10th. Since January 1st, 5·20 inches of rain have fallen on, however, only 28 days.

MARCH.—This was a tolerably favourable month. North-westerly winds preponderated as in February, and while the barometer was often high off the W. and S.W. of Ireland, atmospherical depressions passed south-eastwards across Scandinavia and the North Sea, producing cold N.W. winds and frequent showers. The beginning of the month was cold, but some genial weather was experienced from time to time.

In Dublin the mean temperature (44·0°) was above the average (42·3°); the mean dry bulb readings at 9 a.m. and 9 p.m. were 43·2°. In the twenty-four years ending with 1888, March was coldest in 1867 and 1883 (M. T. = 39·0°) and warmest in 1859 (M. T. = 47·6°). In 1879, the M. T. was 42·3°, in the year 1879 (the cold year), it was 42·9°, in 1888 it was as low as 39·8°.

As a general rule, February in Dublin is only a shade colder than March. This is due to the fact that the Continental anticyclone embraces the British Isles and Scandinavia in March, causing easterly winds. In the present year, however, February was 3·7° colder than March.

The mean height of the barometer was 29·939 inches, or 0·076 inch above the average value for March—namely, 29·863 inches. The mercury rose to 30·548 inches at 9 p.m. of the 18th, and fell to 28·944 inches at 9 p.m. of the 10th. The observed range of atmospherical pressure was, therefore, as much as 1·604 inches—that is, more than an inch and a half. The mean temperature deduced from daily readings of the dry bulb thermometer at 9 a.m. and 9 p.m. was 43·3°, or 3·7° above the value for February, 1889; that calculated by Kaemtz's formula from the means of the daily maxima and minima was 43·0°, or 0·9° above the average mean temperature for March, calculated in the same way, in the twenty years, 1865–84, inclusive (42·1°). The arithmetical mean of the maximal and minimal readings was 44·0°, compared with a twenty-three years' average of 42·3°. On the 29th, the thermometer in the screen rose to 55·3°—wind W.S.W.; on the 3rd, the temperature fell to 31·0°—wind, S.E. The minimum on the grass was 24·2° on the 27th. The rainfall was only 1·676 inches, distributed, however, over 17 days. The average rainfall for March in the twenty-three years, 1865–87, inclusive, was 2·090 inches, and the average number of rainy days was 16·8. The rainfall, therefore, was much below the average, while the rainy days were slightly above the average.

In 1867 the rainfall in March was very large—4·972 inches on 22 days; in 1881, 3·763 inches fell on 18 days; in 1888 also 3·629 inches fell on 23 days. On the other hand, in 1871, only ·815 of an inch was measured on 13 days; and in 1874 only ·958 inch fell, also on 13 days. In 1887 (the "dry year"), 1·408 inches of rain fell on 15 days. A solar halo appeared on the 6th. The atmosphere was foggy on the 1st, 2nd, 13th, and 27th. High winds were noted on 9 days, reaching the force of a gale on 4 days—the 10th, 23rd, 24th, and the 25th. Snow or sleet occurred on the 1st and 2nd; and hail fell on the 1st, 2nd, 5th, 21st, and 31st. The temperature exceeded 50° in the screen on 19 days, compared with 3 days in February, and the same number of days in January; while it fell to or below 32° in the screen on 5 days, compared with only 4 in February and only 3 in January, but with 11 in March, 1888. The minima on the grass were 32°, or less, on 14 nights, compared with 21 nights in February and 16 nights in January. On 2 days the thermometer did not rise to 40° in the screen.

The month opened with very severe weather. At 8 a.m. of Saturday, the 2nd, the thermometer stood at 20° at Aberdeen, and 22° at Parsonstown. The reading in Dublin at this time was 32°.

The weather of the week ending Saturday, the 9th, was very changeable and for the most part cold. At the beginning a keen S.E. wind blew, with hazy, dry weather. On Monday, the 4th, a slow drizzling rain prevailed, while temperature did not exceed 40·1°. The two following days were chiefly fine and much milder, the thermometer reaching a maximum of 50·9° on Wednesday.

On this day the barometer gave way suddenly, and next morning two areas of low pressure were found—one off the north of Scotland, the other over St. George's Channel. The latter system caused the wind to shift to N.E. and N. on Thursday, with dull, cool weather in Ireland. Friday, the 8th, was cloudy but fine—a canopy of cirro-stratus moving from S.W. in an upper current, while the surface wind remained northerly. Saturday was bright, with some passing showers in the afternoon. In Great Britain very severe weather was experienced during the week. At first some intense frosts occurred over a snow-covered country—thus, the thermometer in the screen sank on Sunday to 14° at Cambridge, on Monday to 15° at Shields, and on Tuesday to 19° at Aberdeen. Afterwards, on Thursday and Friday, the depression above mentioned caused drenching rains or heavy falls of sleet and snow in England. In Dublin the mean height of the barometer was 29·788 inches—pressure ranging from 30·013 inches at 9 p.m. of Tuesday (wind, W.S.W.) to 29·389 inches at 9 p.m. of Friday (wind, N.N.W.). The mean dry bulb temperature at 9 a.m. and 9 p.m. was 39·9°. The arithmetical mean of the highest and lowest daily temperatures was 39·6°. Temperature in the screen rose to 50·4° on Wednesday (wind, S.W.), having fallen to 31·0° on Sunday (wind, S.E.). The rainfall amounted to ·105 inch, and was distributed over four days. The largest daily precipitation was ·047 inch on Monday.

Very favourable weather was experienced during the second week (10th to the 16th, inclusive), particularly in Ireland, where temperature was high and the air mild after Monday, the 11th. On this day an area of high atmospherical pressure showed itself over Ireland, and this system held for several days, producing dry weather and N.W. winds. On Wednesday an area of low pressure passed southeastwards across the S. of Scandinavia and Denmark, causing snow and frost in those countries and low temperatures with cold showers in Great Britain also. At the close of the week the anticyclone was giving way in Ireland, where the wind backed towards W. with rising temperature and increasing cloud. In Dublin the mean height of the barometer was 30·214 inches, pressure increasing from 29·762 inches at 9 a.m. of Sunday (wind, W.) to 30·544 inches at 9 p.m. of Friday (wind, S.S.E.). The mean temperature was 46·3°, or 6·3° above the value for the previous week. The highest reading of the screened thermometers was 57·7° on Wednesday (wind, N.W.); the lowest was 31·7° on Monday (wind, W.N.W.). The mean dry bulb temperature at 9 a.m. and 9 p.m. was 46·3°. Only a few light showers fell during the week, yielding ·020 inch of water. Of this quantity ·010 inch was measured on Monday and the same amount on Thursday. Except on Sunday and Wednesday, the percentage of cloud was high. Friday night was very cold in the S.E. of England, and severe frost prevailed in Germany and France on Saturday—at Munich the temperature at 8 a.m. was only 13°.

The chief feature in the weather-record of the third week was the passage across the British Islands of a very deep depression in the interval between Monday, the 18th, and Thursday, the 21st. At the beginning of the week the barometer was high and the weather was fine. During Sunday and Monday a great and uniform reduction of pressure took place, the wind backed to S.W. and S., with light showers and rising temperature. On Tuesday a well-marked depression, which hourly grew deeper, spread eastwards across Ireland and the Irish Sea to the North of England, where the barometer fell below 28·80 inches on Wednesday morning. By this time the wind had shifted to N. and N.E. in Ireland, and rose to a fresh or strong gale. The depression now changed its course, travelling southwards to France, where it "filled up" on Friday. A great deal of cold rain, wet snow, and hail fell in Great Britain in connexion with this disturbance. During Friday and Saturday the wind once more backed to W.S.W., and freshened with a rainy appearance and higher temperature. In Dublin the mean height of the barometer was 29·746 inches—pressure decreasing from 30·301 inches at 9 a.m. of Sunday (wind, W.N.W.) to 28·944 inches at 9 p.m. of Tuesday (wind, W.). The mean dry bulb temperature at 9 a.m. and 9 p.m. was 43·6°, while the mean of the daily highest and lowest readings was 43·8°. Temperature rose to 51·0° in the screen on Saturday (wind, W.S.W.), having fallen to 32·3° on Friday (wind, W.). Rain was measured on three days—the total precipitation amounting to ·571 inch, of which ·221 inch was caught on Monday, and ·331 inch on Tuesday. A little hail fell on Thursday morning.

Fresh N.W. winds, cloudy skies, and chiefly mild weather characterized the week ending Saturday, the 30th. The barometer was throughout high in the S. and S.W., but comparatively low in the far N. On Wednesday an anticyclone, in which the barometer rose above 30·40 inches, advanced over Ireland, where a brief frost occurred. Next day this system had travelled southeastwards to Brittany and the mouth of the English Channel, in which regions the barometer stood above 30·50 inches. Temperature rose fast in Ireland, as a squally westerly wind spread over the country, and rain fell in considerable quantities. The last two days were very mild, but changeable and rather cloudy. On Saturday evening the sky cleared and temperature gave way quickly. Venus shone brightly in the western sky on several evenings. In Dublin the mean height of the barometer was 30·286 inches, pressure increasing from 30·069 inches at 9 a.m. of Sunday (wind, W.N.W.) to 30·441 inches at 9 a.m. of Wednesday (wind, also W.N.W.). The mean dry bulb temperature at 9 a.m. and 9 p.m. was 47·3°, while the arithmetical mean of the daily maxima and minima was 48·1°. The thermometer in the screen rose to 55·3° on Thursday (wind, W.S.W.), having fallen to 31·2° on Wednesday (wind, N.W.). Rain fell on five days to the amount of ·190 inch; the maximal fall in 24 hours was ·105 inch on Wednesday, or rather early on Thursday morning.

The last day of the month—Sunday, the 31st—was squally, cold, and showery. A sharp hail-shower fell at 2 p.m.

The rainfall in Dublin during the three months ending March 31st has amounted to 5·785 inches on 43 days, compared with 5·087 inches on 43 days during the same period in 1888, and a 24 years' average of 6·494 inches on 51·4 days. While double as much rain as fell in January and February of last year fell in the same months this year, the rainfall in March of this year is not one-third that of March, 1888—namely, 0·753 inches.

At Greystones, Co. Wicklow, the rainfall in March, 1889, was 1·71 inches, distributed over only 7 days. Of this quantity ·61 inch fell on the 18th. Since January 1st, 7·97 inches of rain have fallen on, however, only 38 days.

April.—Like April, 1888, this was a generally cold, cloudy, changeable, showery month. It showed a low atmospherical pressure, and a preponderance of winds from polar points of the compass. In Dublin rain fell on as many as 21 days, and the total amount was more than 2½ inches. With the exception of 1879, 1881, 1887, and 1888, this was the coldest April experienced in Dublin since the present records were commenced in 1865.

In Dublin the mean temperature (46·1°) was decidedly below the average (47·5°); the mean dry bulb readings at 9 a.m. and 9 p.m. were 45·6°. In the twenty-four years ending with 1888 April was coldest in 1879 (the cold year) (M. T. = 44·6°) and warmest in 1865 and 1874 (M. T. = 50·4°). In 1886, the M. T. was 46·3°, in 1887 it was as low as 44·1°, and in 1888 it was only 44·7°.

The mean height of the barometer was 29·748 inches, or 0·110 inch below the average value for April—namely, 29·857 inches. The mercury rose to 30·153 inches at 9 a.m. of the 18th, and fell to 29·179 inches at 9 a.m. of the 4th. The observed range of atmospherical pressure was, therefore, only 0·974 inch—that is, a little less than one inch. The mean temperature deduced from daily readings of the dry bulb thermometer at 9 a.m. and 9 p.m. was [illegible], or only [illegible] above the value for March, 1889; that calculated by Kaemtz's formula from the means of the daily maxima and minima was [illegible], or [illegible] below the average mean temperature for April, calculated in the same way, in the twenty years, 1865–84, inclusive ([illegible]). The arithmetical mean of the maximal and minimal readings was [illegible], compared with a twenty-three years' (1865–1887, inclusive) average of [illegible]. On the [illegible], the thermometer in the screen rose to [illegible]—wind W.; on the [illegible] the temperature fell to [illegible]—wind N.W. The minimum on the grass was [illegible] on the same date. The rainfall was [illegible] inches, distributed over 21 days. The average rainfall for April in the twenty-three years, 1865–87, inclusive, was [illegible] inches, and the average number of rainy days was [illegible]. The rainfall, therefore, was considerably above the average, while the rainy days much exceeded it. In 1877 the rainfall in April was very large—[illegible] inches on 21 days; in [illegible] also [illegible] inches fell on [illegible] days. On the other hand, in 1873, only [illegible] of an inch was measured on [illegible] days, and in 1870 only [illegible] of an inch fell, also on 5 days.

No solar halos were seen. The atmosphere was more or less foggy on the [illegible], [illegible], and [illegible]. High winds were noted on 10 days, but on only two occasions was the force of a gale attained—namely, on the [illegible] and 20th. Snow or sleet occurred on the 4th, and hail fell on the [illegible], [illegible], [illegible], and [illegible]. The temperature exceeded 50° in the screen on 19 days, compared with 13 days in March, 8 days in February, and the same number of days in January. It did not fall to 32° in the screen on even one night. The minima on the grass were 32°, or less, on [illegible] nights, compared with 13 nights in March, [illegible] in February, and 16 nights in January. The mean lowest temperature on the grass was [illegible], compared with [illegible] in 1888 and [illegible] in 1887. A peal of thunder was heard to the southward about [illegible] p.m. of the [illegible].

Unsettled weather prevailed during the greater part of the period ending Saturday, the 6th. Temperature showed a marked decrease, and squally N.W. winds and cold showers were general. At the beginning of the period a depression was travelling southeastwards across the south of Scandinavia. Strong N.W. winds and showers of cold rain, sleet and hail were reported from many stations on Monday, the 1st. In Scotland Tuesday was a very inclement day. The barometer now gave way quickly, and a deep depression advanced from the N.W. over Ireland and England. It caused heavy rains in the N. of Ireland, and all over the country the weather became broken, gloomy, cold, and showery, continuing so until the end of the week. In Dublin the mean height of the barometer was [illegible] inches—pressure ranging from [illegible] inches at 9 a.m. of Tuesday (wind, N.N.W.) to [illegible] inches at 9 a.m. of Thursday (wind also N.N.W.). The mean dry bulb temperature at 9 a.m. and 9 p.m. was [illegible]. The arithmetical mean of the daily maxima and minima was almost the same—namely, [illegible]. The screened thermometers rose to [illegible] on Monday (wind, N.W.), and fell to [illegible] on Thursday (wind, N.N.W.). Rain was measured on three days, the total amount being [illegible] inch, of which [illegible] inch was registered on Thursday. Some sleet fell on this day. There were thunderstorms over the north of England on Friday, the 5th.

Many years have passed since so dull and wet and cold a period as that ended Saturday, the 13th, was experienced in April. A continuously low barometer for many days, chilly polar winds, densely overcast skies, frequent and abundant rainfalls, and an extremely low temperature were the characteristics of this week. As usually happens when N.E. winds are prevalent, the weather was much worse on the east coast of Ireland than in the centre and western parts of this country. At 8 a.m. of Sunday, the 7th, the centre of a well-defined atmospherical depression was found off Ushant, so that easterly and north-easterly winds prevailed in the United Kingdom. In Dublin this day was very wet and cold. A short spell of sunshine on Monday morning was followed by a dull damp day. Tuesday was raw, gloomy, and damp; Wednesday showed a slight improvement; but Thursday was again cold and wet. On Friday the weather improved; and Saturday was dry, but cold. In Dublin the mean height of the barometer was only [illegible] inches—pressure ranging from [illegible] inches at 9 p.m. of Monday (wind, N.W.) to [illegible] inches at 9 p.m. of Saturday (wind, N.E.). The mean temperature deduced from readings of the dry bulb thermometers taken daily at 9 a.m. and 9 p.m., was [illegible]. The arithmetical mean of the highest and lowest daily temperatures was [illegible]. The shade temperature was highest ([illegible]) on Friday (wind, N.E.), and lowest ([illegible]) on Monday (wind, E.). Rain fell on six days to the amount of [illegible] inches—the maximal fall in 24 hours being [illegible] inch, on Thursday. The atmosphere was foggy on Wednesday and Friday.

Favourable, dry weather prevailed during the third week ([illegible] inclusive). At first the air was dry and searching by day, frosty by night; but on Wednesday, the 17th, it became milder, and so continued to the end. From Sunday until Thursday an area of comparatively high atmospherical pressure was found over the Atlantic to the westward of Ireland, where the winds were, accordingly, northerly (N.E. to N.W.), and the weather was fine and dry. In Dublin sharp ground frosts occurred on each of the first three nights—the grass minima being [illegible], [illegible], and [illegible], respectively. On Thursday the anticyclone (area of high pressure) moved southwards, and a warm southwesterly air current spread over Ireland and Scotland, causing higher temperatures and increasing cloud. On Friday night rain fell heavily in the N.W. of Ireland, but only a shower or two passed over Dublin. In this city the mean height of the barometer was [illegible] inches, pressure ranging between [illegible] inches at 9 a.m. of Thursday (wind, S.S.W.) and [illegible] inches at 9 p.m. of Saturday (wind, W.S.W.). The mean dry bulb temperature at 9 a.m. and 9 p.m. was [illegible], or [illegible] above the corresponding value for the previous week. The arithmetical mean of the highest and lowest daily temperatures was [illegible]. The thermometer rose nearly to 60° in the shade (screen) on Wednesday, Thursday, and Saturday—the actual values being [illegible], [illegible], and [illegible] respectively. On Monday the minimum in the shade was [illegible], while that on the grass was as low as [illegible]. Rain fell on the night of Good-Friday, the 19th, to the amount of only [illegible] inch, and on Saturday night to the amount of [illegible] inch—total rainfall for the week = [illegible] inch, on two days.

Conditions were cyclonic over North-western Europe throughout the fourth week (21st–27th inclusive), and consequently the weather was unsettled, squally, and showery, with some fine intervals. Temperature ruled decidedly low for the time of year, and in Dublin hail fell on Sunday and the three following days. On Easter Day, April 21, the centre of an extensive atmospherical depression lay off the extreme N.W. of Scotland—the barometer at 8 a.m. ranged from 30·22 inches in the S.E. of France to 29·17 inches at Stornoway (Hebrides). Showers of rain and hail and squalls prevailed. On Tuesday another depression travelled southeastwards down the North Channel and across the Irish Sea to the centre and east of England. Heavy showers fell in all districts and thunder and lightning with hail occurred at many places in England. On Friday a new V-shaped depression caused a heavy rainfall in Dublin. Towards the close of the week a tendency for southerly to take the place of northerly winds was perceptible. In Dublin the mean height of the barometer was 29·741 inches—pressure ranging from 29·384 inches at 7 p.m. of Tuesday (wind, W.), to 30·016 inches at 9 p.m. of Thursday (wind, S.E.). The mean dry bulb temperature at 9 a.m. and 9 p.m. was 45·1°. The mean temperature derived from the highest and lowest daily readings was 46·5°. The screened thermometers rose to 54·7° on Tuesday (wind, W.), and fell to 35·1° on Monday (wind, N.N.W.). Rain fell on seven days to the total amount of ·838 inch. The maximal rainfall in 24 hours was ·372 inch on Friday (wind, S. to W.N.W.).

On Sunday, the 28th, a S.E. gale blew for some hours in Dublin, while great cold prevailed inland, and sleet and snow fell. Heavy showers fell on the 29th, and on the 30th there was another short-lived southeasterly to easterly gale. At 9 p.m. distant thunder was heard near Dublin, and great gloom occurred for a time.

The rainfall in Dublin during the four months ending April 30th has amounted to 8·443 inches on 74 days, compared with 9·090 inches on 88 days during the same period in 1888, 5·637 inches on 52 days during the same period in 1887, and a 23 years' average of 9·458 inches, on 68·2 days.

At Greystones, Co. Wicklow, the rainfall in April, 1889, was 2·01 inches, distributed over only 12 days. Of this quantity ·45 inch fell on the 11th, and ·41 inch on the 5th. Since January 1st, 7·98 inches of rain have fallen at Greystones, on, however, only 41 days.

May.—This was a very favourable month—singularly free from prevailing easterly winds, with a high mean temperature, and [illegible] rainfall. The atmosphere was generally in a quiet state, no deep depressions and no gales being observed. In Dublin calms were of frequent occurrence, and in consequence morning fogs formed over the city.

In Dublin the mean temperature (54·6°) was much above the average (51·9°); the mean dry bulb readings at 9 a.m. and 9 p.m. were 54·1°. In the twenty-four years ending with 1888, May was coldest in 1869 (M. T. = 48·2°), in 1885 (M. T. = 48·7°), and in 1879 (the "cold year") M. T. = 48·3°. It was warmest in 1868 (the "warm year") (M. T. = 55·8°) and 1875 (M. T. = 54·9°). In 1886, the M. T. was 50·8°; in 1887 it was 51·4°, and in 1888 it was 52·6°.

The mean height of the barometer was 29·790 inches, or 0·216 inch below the average value for May—namely, 29·996 inches. The mercury rose to 30·113 inches at 9 p.m. of the 26th, and fell to 29·440 inches at 3 p.m. of the 31st. The observed range of atmospherical pressure was, therefore, only 0·673 of an inch—that is, a little less than seven-tenths of an inch. The mean temperature deduced from daily readings of the dry bulb thermometer at 9 a.m. and 9 p.m. was 54·1°, or 8·5° above the value for April, 1889; that calculated by Kaemtz's formula from the means of the daily maxima and minima was 53·4°, or 2·5° above the average mean temperature for May, calculated in the same way, in the twenty years, 1865–84, inclusive (50·9°). The arithmetical mean of the maximal and minimal readings was 54·6°, compared with a twenty-three years' average of 51·9°. On the 31st, the thermometer in the screen rose to 72·8°—wind, E.; on the 11th the temperature fell to 40·8°—wind calm. The minimum on the grass was 35·3° on this same date. The rainfall amounted to 2·131 inches, distributed over 17 days. The average rainfall for May in the twenty-three years, 1865–87, inclusive, was 2·073 inches, and the average number of rainy days was 15·5. The rainfall and the rainy days, therefore, were somewhat above the average. In 1886 the rainfall in May was very large—5·477 inches on 21 days; in 1869 also 5·414 inches fell on 19 days. On the other hand, in 1871, only ·378 of an inch was measured on 8 days; in 1876 only ·798 of an inch fell on 6 days; in 1887 only ·883 of an inch fell on 10 days; and in 1888 only ·978 of an inch on 11 days.

There were solar halos on the 10th and 30th. The atmosphere was foggy on the 5th, 6th, 11th, 12th, 14th, 16th, 19th, 21st, and early morning of the 27th. High winds were noted on only 4 days, attaining the force of a gale on not one occasion. Hail occurred on the 31st. The only heavy falls of rain were on the 5th (·503 inch), and on the 12th (·453 inch). There were no electrical disturbances in Dublin.

The first four days were changeable, with a damp atmosphere and cloudy sky on Friday and Saturday, the 3rd and 4th.

During the week ended Saturday, the 11th, the weather was again changeable, with some heavy rainfalls in Ireland, but temperature showed a decided advance. Sunday and Thursday were very wet days in Dublin, while Monday, Tuesday, and Saturday were equally fine. Rain fell for many hours on Sunday, and in the evening a severe thunderstorm occurred in Ulster. A dense vapour fog next morning gave place to brilliant sunshine. Then came two tolerably fine, but not settled, days—to be followed by a very wet day (Thursday). Friday was at first squally with a fine drizzle, but the afternoon proved fair, and the following night was cold. Saturday was a very fine day, with lofty cirrus from S.E. In Dublin the mean height of the barometer was 29·768 inches—pressure receding to 29·484 inches at 9 p.m. of Thursday (wind E.S.E.), and increasing to 29·912 inches at 9 p.m. of Saturday (wind, light from E.). The mean dry bulb temperature at 9 a.m. and 9 p.m. was 53·8°. The mean of the daily maxima and minima in the screen was 55·2°. On

Tuesday the shade thermometer rose to 69·6° (wind, W.S.W.). This was the first occasion this season upon which temperature in Dublin exceeded 60° in the shade. On Saturday the minimum in the shade was 40·5°, and on the grass it was 31·9° (wind, calm). Rain was measured on five days, the total amount being 1·081 inches, of which ·509 inch was registered on Sunday (wind, S. to N.W.). Very high temperatures (70°—77°), were recorded in central and eastern England during this week. On Friday an area of low pressure approached England from Germany, causing a heavy rainfall.

Cloudy, dull, quiet weather prevailed during the period from the 12th to the 18th inclusive. At the beginning of the week a cyclone in which the gradients, however, were not steep, was found with its centre over Norfolk, whence it had travelled slowly from Germany. Much rain fell in eastern and central England in connection with this system, and even in Dublin ·[illegible] inch of rain was measured on Monday morning. This depression gradually filled up and dispersed, when atmospherical pressure became so uniform over Western Europe that on Tuesday morning (the 14th) the barometer readings over the whole of our islands, the North Sea, the Netherlands, Denmark, the South of Scandinavia, France and Germany were all within one-tenth of an inch of each other. Consequently, very light, variable winds and quiet weather prevailed. On the afternoon of this day a severe thunderstorm occurred in the north-east of Ireland and south-west of Scotland. A solar halo appeared in Dublin on Thursday afternoon, and Friday proved a damp, dropping day. Saturday was at first cloudy, but afterwards bright and warm. In Dublin the mean height of the barometer was 29·910 inches, pressure ranging only from 29·780 inches at 9 p.m. of Tuesday (wind, calm), to 29·[illegible] inches at 9 a.m. of Friday (wind, S.S.E., light). The mean dry bulb temperature at 9 a.m. and 9 p.m. was 58·1°, while the mean daily highest and lowest readings of the screened thermometers were 58·[illegible]°. The highest shade temperature was 69·6° on Saturday (wind, variable); the lowest was [illegible]° on Tuesday (wind, E.N.E.). Rain fell on three days to the amount of ·[illegible] inch,—the heaviest rainfall in 24 hours being ·[illegible] inch on Sunday (wind, W.S.W. to N.W.).

Favourable weather held throughout the week ending Saturday, the 25th, in Dublin, where only light showers fell on Friday. Until that day temperature ruled unusually high for the time of year —the daily maxima in the screen being 66·3°, 69·6°, 72·5°, 68·6°, and 67·3° respectively. On Wednesday a gradual fall of the barometer set in and continued until Friday morning, when pressure fell to 29·5 inches in the Hebrides and 29·[illegible] inches at Roche's Point, Co. Cork. On Thursday severe thunderstorms were reported from the east of England, and at night heavy rain set in over the north and north-west of Ireland. On this day temperature rose to 82° in London and to 81° at Loughborough and Cambridge. Early on Friday a rapid fall of temperature occurred in Ireland, so that in Dublin the thermometer did not exceed 56·9° on this day, while it fell to 44·1° in the screen and to 41·9° on the grass during the ensuing night. The mean height of the barometer was 29·[illegible] inches—pressure rising to 30·148 inches at 9 p.m. of Monday (wind, E.), and falling to 29·[illegible] inches at 7 a.m. of Friday (wind, S.W.). The mean dry bulb temperature at 9 a.m. and 9 p.m. was 58·8°, while the mean of the highest and lowest daily readings was 58·3°. The screened thermometers rose to 72·5° on Tuesday (wind, E.N.E.), and fell to 44·1° on Saturday (wind, N.N.W.). Rain fell on Friday to the amount of only ·010 inch.

Throughout the closing period (Sunday, 26th, to Friday, 31st) much less settled weather prevailed. At first the barometer was lowest over the Continent, and large quantities of rain fell in England, while the weather was comparatively fine in Ireland. From Tuesday, the 28th, however the lowest pressures were found off the north-west and west of Ireland, so that fresh to strong southerly (S.E. to S.W.) winds and frequent showers were experienced. Thunder and lightning accompanied the showers in several places, but—as often happens—Dublin escaped the electrical disturbances. During this period, the barometer ranged between 30·[illegible] inches at 9 a.m. of Monday (wind, E.) and 29·[illegible] inches at 5 p.m. of Tuesday (wind, S.). On Monday, the 27th, the thermometer rose to 64·6° in the shade (wind, E.S.E.), the minimum during the previous night having been 46·[illegible]°.

During the month the thermometer in the screen did not fall below 40·5° (on the 11th); and on only one night was a temperature of 32° or less recorded on the grass. The mean minimal temperature on the grass was 42·4°, compared with 37·3° in May, 1888, and 37·9° in May, 1887.

The rainfall in Dublin during the five months ending May 31st has amounted to 10·[illegible] inches on [illegible] days, compared with 9·[illegible] inches on [illegible] days during the same period in 1888, [illegible] inches on [illegible] days in 1887, and a 23 years' average of 10·[illegible] inches, on 8[illegible] days.

At Greystones, Co. Wicklow, the rainfall in May, 1889, was 1·[illegible] inches, distributed over only 10 days. Of this quantity ·[illegible] inch fell on the 3rd, ·47 inch on the 4th, and ·[illegible] inch on the 25th. Since January 1st, 13·[illegible] inches of rain have fallen at Greystones, on, however, only 81 days.

June.—This month shows a marked resemblance to June, 1887, which will be long remembered as one of the driest and warmest months on record in Dublin. The mean temperature fell indeed [illegible]° short of that of June, 1887, but it was still 1·7° above the average; the mean height of the barometer also was ·[illegible] inch lower, but it was ·153 inch above the average for the month; and there was even a greater deficiency of rainfall than had occurred in June, 1887. Lastly, there was again a preponderance of easterly and north-easterly winds, with the same [illegible] fine summer weather which was so characteristic of the Queen's Jubilee summer.

In Dublin the mean temperature (59·5°) was decidedly above the average (57·8°); the mean dry bulb readings at 9 a.m. and 9 p.m. were 60·[illegible]°. In the twenty-four years ending with 1888, June was coldest in 1882 (M. T. = 55·[illegible]°) and in 1879 (the "cold year") (M. T. = [illegible]°). It was warmest in 1887 (M. T. = 62·[illegible]°), in 1868 (M. T. = 61·0°), and in 1865 (the "warm year") (M. T. = [illegible]°). In 1888 the M. T. was 57·[illegible]°, and in 1885 it was 56·2°.

The mean height of the barometer was 30·077 inches or 0·153 inch above the average value for June—namely, 29·924 inches. The mercury marked 30·482 inches at 9 a.m. of the 5th, and fell to

30·051 inches at 9 a.m. of the 2nd. The observed range of atmospheric pressure was, therefore, ·[illegible] of an inch—that is, a little more than eight-tenths of an inch. The mean temperature deduced from daily readings of the dry bulb thermometer at 9 a.m. and 9 p.m. was [illegible], or [illegible] above the value for May, 1889; that calculated by Kaemtz's formula from the means of the daily maxima and minima was [illegible], or [illegible] above the average mean temperature for June, calculated in the same way, in the twenty years, 1865–84, inclusive ([illegible]). The arithmetical mean of the maximal and minimal readings was [illegible], compared with a twenty-three years' average of 57·8°. On the 22nd, the thermometer in the screen rose to [illegible]—wind E.N.E.; on the 17th the temperature fell to 41·2°—wind N.E. The minimum on the grass was [illegible] on the latter date. The rainfall amounted to only one-tenth of an inch (·103 inch), distributed over 6 days. There was no measurable rainfall in Dublin after the 14th, and not since September, 1865—when only ·054 inch was registered on three days—has so comparatively rainless a month occurred at this station. The average rainfall for June in the twenty-three years, 1865–87, inclusive, was [illegible] inches, and the average number of rainy days was [illegible]. The rainfall and the rainy days, therefore, were remarkably below the average. In 1878 the rainfall in June was very large—[illegible] inches on 18 days; in 1879 also [illegible] inches fell on 24 days. On the other hand, in 1874, only ·[illegible] of an inch was measured on 9 days; and in 1868 only ·077 of an inch fell on but 6 days.

In 1887 the rainfall was only ·[illegible] of an inch, distributed over only 5 days; in 1888 it was as much as 3·046 inches, distributed over as many as 18 days.

High winds were noted on 2 days, but did not attain the force of a gale on any occasion. Temperature reached or exceeded 70° in the screen on 10 days, as compared with 11 days in 1887, and only one day in 1888. Luminous cloud were seen from Dublin on the northern horizon on the night of the 7th. There were no electrical disturbances in or near Dublin.

The first day of the month was cloudy, squally, and showery.

So far as Dublin is concerned, beautiful weather was experienced in the week ending Saturday, the 8th. Only on the afternoon of Monday and Saturday were conditions unsettled and slight rain fell. The amount of cloud at other times was small, and much brilliant sunshine was enjoyed. At the beginning of the week an extensive depression, which had caused a fresh gale and very heavy rain in the S. and W. of Ireland on Saturday the 1st inst., was passing away over the Atlantic. Temperature rose on Sunday the 2nd, to 83° in London, 85° at Cambridge, and 84° at Loughborough, and in the afternoon violent thunderstorms, with hail, occurred in many parts of Great Britain. An area of high pressure now advanced northwards over the British Islands from France and beautiful weather set in, lasting until Thursday. Towards night on this day a small depression developed over the western part of the English Channel and severe thunderstorms again became prevalent in England. On Saturday another depression appeared in the N.W., causing some rain in Dublin. The approach of this system was heralded on Friday night by the appearance of luminous clouds on the northern horizon. In Dublin the mean pressure was [illegible] inches—highest [illegible] inches at 9 a.m. of Wednesday (wind, E.); lowest, [illegible] inches at 9 a.m. of Sunday (wind, S.E.). The mean dry bulb temperature at 9 a.m. and 9 p.m. was [illegible]. The mean of the maxima and minima combined was [illegible]. The shade thermometers rose to 71·8° on Thursday (wind, E.), and fell to [illegible] on Saturday (wind, W.N.W.). Rain fell on two days to the amount of ·[illegible] inch, of which ·[illegible] inch was registered on Saturday.

During the second week (9th–15th, inclusive), strange contrasts of weather were experienced in different parts of the United Kingdom. Thus, while clear skies and brilliant sunshine obtained in Ireland during the first three days, over the greater part of England there were gloomy skies, low temperature, strong N.E. winds, and heavy rains. Whitsun Monday, the 10th, which was a cloudless day in Dublin, was cold and wet as a day in midwinter in London. This singular contrast was brought about by the presence of a trough of low barometrical pressure over the north of France, Holland, and Denmark, while the barometer was high and rising in Ireland and Scotland. After Tuesday this distribution of pressure gave place to irregular and slight gradients for S.W. and W. to N.W. winds, so that the weather became less settled in Ireland, but improved in England. In Dublin the mean height of the barometer during the week was [illegible] inches, pressure ranging between [illegible] inches at 9 p.m. of Thursday (wind, S.S.W.) and [illegible] inches at 9 p.m. of Saturday (wind light from N.). The mean dry bulb temperature at 9 a.m. and 9 p.m. was [illegible], while the mean of the daily maxima and minima was 57·7°. Temperature in the screen rose to [illegible] on Thursday (wind southerly), having fallen as low as [illegible] on Tuesday morning (wind, N.E.). The rainfall was only ·048 inch, distributed over three days. Since June 1 only ·100 inch of rain had fallen in Dublin, on five days. Up to Saturday, the 15th, [illegible] inches had been measured in London, on however only five days.

Quiet, summerlike weather prevailed in Ireland throughout the third week (16th–22nd, inclusive). Less settled conditions, with fresh N.E. winds, and at times a good deal of cloud, were reported from Great Britain. All through, an anticyclone of moderate intensity held over the British Islands, while areas of relatively low pressure appeared from time to time over the Baltic and Germany and France. The first three and the last two days were almost cloudless in Dublin, although the atmosphere was hazy. During the earlier part of Wednesday and Thursday the sky was overcast, but in the afternoons of both these days the cloud canopy broke up. There were some heavy thunderstorms in France each day after Monday. In Dublin the mean height of the barometer was 30·190 inches, pressure ranging only from [illegible] inches at 9 a.m. of Tuesday (wind, E.), to 30·118 inches at 9 p.m. of Thursday (wind, calm). The mean dry bulb temperature at 9 a.m. and 9 p.m. was [illegible], while the mean daily maxima and minima combined was [illegible]. The screened thermometers rose to 74·9° on Saturday, having fallen to 46·0° on Sunday. There was no rain during the week. The wind varied between E. and N., except sometimes at night when a light westerly land breeze was felt in Dublin. On Wednesday and Thursday the easterly wind freshened considerably.

The period from the 23rd to the 29th inclusive was another week of quiet, summer-like weather. After Tuesday, the easterly breezes, which had been so prevalent, gave place to light and variable, or moderate westerly winds. Barometrical gradients were slight and irregular throughout. On Tuesday and the two following days thunderstorms occurred in different parts of France. In the British Islands only a few showers fell during the week. After Tuesday daily maxima of 80° or upwards were recorded at several English stations. In Ireland the thermometer rose (at Parsonstown) to 77° on Sunday. In Dublin the mean height of the barometer was 30·123 inches, pressure ranging from 29·944 inches, at 9 a.m. of Monday (wind, E.), to 30·273 inches at 9 a.m. of Saturday (wind, N.W.). The mean dry bulb temperature at 9 a.m. and 9 p.m. was 62·2°. The mean temperature deduced from the daily highest and lowest readings was 62·4°. The shade maxima exceeded 70° on five days—the absolute maximum being 72·9° on Wednesday (wind, W.N.W.). The minimum for the week was 52·1° on Tuesday (wind, calm). A few large "spots" of rain were observed about 5 p.m. of Wednesday, and at 8 p.m. of Saturday a light shower fell, yielding only three-thousandths of an inch of water—less than the standard minimal rainfall for measurement.

Sunday, the 30th, was at first cloudy and then fair. In the afternoon the air was very dry, and a peculiar yellowish-grey haze overspread the sky.

The rainfall in Dublin during the six months ending June 30 has amounted to 10·478 inches on 97 days, compared with 12·113 inches on 97 days during the same period in 1888, 8·741 inches on 67 days in 1887, and a 23 years' average of 12·059 inches, on 95·7 days.

At Greystones, Co. Wicklow, the rainfall in June, 1889, was ·23 inch, distributed over only 2 days. Of this quantity ·18 inch fell on the 3rd. Since January 1st, 10·26 inches of rain have fallen at Greystones, on, however, only 58 days.

JULY.—Warm and fine at the beginning and close, this month on the whole proved changeable, showery, and cool, with a great preponderance of North-westerly winds—a very unusual state of things in an Irish July. Speaking generally, however, the weather was favourable from both a sanitary and an agricultural point of view.

In Dublin the mean temperature (58·7°) was decidedly below the average (60·5°); the mean dry bulb readings at 9 a.m. and 9 p.m. were 58·3°. In the twenty-four years ending with 1889, July was coldest in 1879 (the "cold year") (M. T. = 57·2°). It was warmest in 1887 (M. T. = 63·7°), and in 1868 (the "warm year") (M. T. = 63·8°). In 1880, the M. T. was 61·0°; in 1885, it was as low as 57·3°. From this, 1887 proves to have been the warmest since the present records commenced, whilst July, 1879, was almost the coldest.

The mean height of the barometer was 29·944 inches or 0·022 inch above the average value for July—namely, 29·922 inches. The mercury reached 30·290 inches at 9 a.m. of the 1st, and fell to 29·195 inches at 9 a.m. of the 21st. The observed range of atmospherical pressure was, therefore, 1·095 inch—that is, a little more than an inch. The mean temperature deduced from daily readings of the dry bulb thermometer at 9 a.m. and 9 p.m. was 58·3°, or 1·7° below the value for June, 1889; that calculated by Kaemtz's formula from the means of the daily maxima and minima was 57·6°, or 2·2° below the average mean temperature for July, calculated in the same way, in the twenty years, 1865–84, inclusive (59·8°). The arithmetical mean of the maximal and minimal readings was 58·7°, compared with a twenty-three years' average of 60·5°. On the 6th, the thermometer in the screen rose to 71·9°—wind, W.N.W.; on the 4th the temperature fell to 44·0°—wind, N.N.E. The minimum on the grass was 38·9° on the latter date. The rainfall was 2·976 inches, distributed over 15 days. The average rainfall for July in the twenty-three years, 1865–87, inclusive, was 2·350 inches, and the average number of rainy days was 17·1. The rainfall, therefore, was somewhat above the average, while the rainy days were below it. In 1880 the rainfall in July was very large—6·087 inches on 24 days; in 1871 also 5·321 inches fell on 26 days. On the other hand, in 1870, only ·856 of an inch was measured on 8 days; in 1869, the fall was only ·760 of an inch on 9 days, and in 1868, only ·741 of an inch fell on but 5 days.

High winds were noted on 6 days, but attained the force of even a moderate gale on only two occasions—viz., the 21st and 27th. Temperature reached or exceeded 70° in the screen on only two days. In July, 1887, temperature reached or exceeded 70° in the screen on no fewer than 17 days. In July, 1888, the maximum was 68·7°. There was a storm of thunder, lightning, and heavy rain and hail on the 18th. Fog was observed on the 9th and 30th.

The drought observed in June continued through the first week, but until Saturday, the 6th—which was a really hot day—temperatures ruled lower than in the previous week. The prevailing distribution of atmospherical pressure was anti-cyclonic over the British Islands, with areas of relatively low barometer over both the Baltic and the Mediterranean. At the beginning and close of the week the wind was northwesterly; but in the interval light to moderate easterly breezes (E. to N.E.) prevailed. Up to and including Wednesday the mornings were cloudy—indeed on Tuesday, the 2nd, slight drizzle was observed at 7·30 a.m. Electric cirri showed on Thursday afternoon; and the atmosphere was generally hazy through the week. On Saturday a remarkable rise of temperature occurred—the maximum, 71·9°, being also the maximum of the present month, so far as it has gone. In Dublin the mean height of the barometer was 30·205 inches, pressure ranging from 30·290 inches at 9 a.m. of Monday, the 1st (wind, N.W.) to 29·996 inches at 9 p.m. of Saturday, the 6th (wind, W.N.W.). The mean dry bulb temperature at 9 a.m. and 9 p.m. was 61·0°. The mean of the daily highest and lowest shade readings of the thermometer was 61·7°. The screened thermometers rose to 77·9° on Saturday, having fallen to 50·1° on Thursday.

Although less settled than before, the weather was most favourable in all respects throughout the second week (7th–13th inclusive). Early on the morning of Sunday, the 7th, sharp showers fell, yielding a rainfall of [illegible] inch—the first recorded since June 14. Rain fell on Tuesday and Wednesday abundantly over the greater part of the United Kingdom, the west and centre of Ireland only escaping the copious showers. On Sunday rather a deep barometrical depression lay over the S. of Norway, and fresh N.W. winds with cool changeable weather prevailed in the British Isles. On Monday morning a new depression appeared off the Land's End, whence it travelled slowly to the S. of Ireland, causing the copious rains already mentioned. On Wednesday this system began to move rapidly in a north-easterly direction, and in 24 hours its centre reached the Christiania Fjord, in Norway. Thus, clear weather now set in, so that Snowdon was seen from the Wicklow Coast on Thursday afternoon. On Friday and Saturday very shallow disturbances led to local showers or rain in parts of England and Ireland. In Dublin the mean height of the barometer was [illegible] inches, pressure ranging from [illegible] inches at 9 a.m. of Wednesday (wind, W.N.W.), to [illegible] inches at 9 p.m. of Friday (wind, E.). The mean dry bulb temperature at 9 a.m. and 9 p.m. was [illegible]°. The mean of the daily highest and lowest temperatures in the screen was [illegible]°. The shade thermometers rose to [illegible]° on Thursday (wind, W.S.W.), having fallen to [illegible]° on Monday (wind, N.N.E.). Rain fell on four days to the amount of ·737 inch. The heaviest fall in 24 hours was ·453 inch on Tuesday.

During the week ended Saturday, the 20th, changeable, cloudy, and cool weather was general over the North-west of Europe. The distribution of atmospherical pressure was cyclonic—systems of low pressure, with secondary shallow depressions, causing heavy showers in most places from time to time. On Sunday, the 14th, the atmosphere was very clear, so that the Mourne and Welsh Mountains were plainly seen from the Co. Wicklow. On Monday and Tuesday heavy showers of rain and hail, with occasional thunder and lightning, prevailed in and about Dublin. Three fine, but cloudy, cool days followed, but on Friday night a depression of considerable intensity brought a fresh rainfall. Saturday was close and showery at first, but in the afternoon the weather improved. In Dublin the mean height of the barometer was [illegible] inches, pressure ranging between [illegible] inches at 9 a.m. of Thursday (wind, N.W.), and [illegible] inches at 1 p.m. of Saturday (wind, W.N.W.). The mean dry bulb temperature at 9 a.m. and 9 p.m. was [illegible]°. The arithmetical mean of the daily highest and lowest shade temperatures was [illegible]°. On Wednesday the thermometer rose to [illegible]°, having fallen during the preceding night to [illegible]°. Rain fell on four days—the total measurement being ·728 inch. The largest rainfall in 24 hours was ·279 inch on Tuesday. Nearly as much (·[illegible] inch) was credited to Friday. Hail, thunder, and lightning were observed on Tuesday.

In the fourth week (21st–27th inclusive), the weather was again very unsettled, but on Thursday the rain ceased, and bright sunshine, with high, bracing, northerly winds took their place. Atmospherical pressure was very unstable—a series of depressions passing eastwards along irregular paths until Saturday, when a system of high barometer became established over the West of Ireland. Very heavy showers fell in and about Dublin on the first three days of the week—the resulting measurement of rain being nearly an inch, including ·[illegible] inch registered on Monday. Wednesday broke dull and wet, but in the afternoon conditions improved. During the last three days of the week very strong northerly (N.W. to N.) winds prevailed, accompanied at first with cloudy, but afterwards with clear, bracing weather. In Dublin the mean height of the barometer was [illegible] inches, pressure having fallen to [illegible] inches at 9 a.m. of Sunday (wind, W.S.W.) and risen to [illegible] inches at 9 p.m. of Saturday (wind, N.W.). The mean dry bulb temperature at 9 a.m. and 9 p.m. was [illegible]°. The mean of the daily maxima and minima was [illegible]°—the highest shade temperature was [illegible]° on Saturday; the lowest was [illegible]° on Thursday. Rain fell on five days to the amount of 1·[illegible] inches. The heaviest downpour in 24 hours was ·[illegible] inch on Monday.

During the last four days, quiet summer-like weather prevailed. On the 28th and 29th there was much fog in the Irish Sea, and on the 31st a considerable increase of temperature took place.

The rainfall in Dublin during the seven months ending July 31st has amounted to 13·[illegible] inches, on 112 days, compared with 7·[illegible] inches on 80 days during the same period in 1887, 15·[illegible] inches on 109 days in 1888, and a twenty-three years' average of 14·[illegible] inches, on 11[illegible] days.

At Greystones, Co. Wicklow, the rainfall in July, 1889, was 2·[illegible] inches, distributed over 13 days. Of this quantity ·[illegible] inch fell on the 9th, ·[illegible] inch on the 10th, and ·[illegible] inch on the 20th. Since January 1, 1[illegible] inches of rain have fallen at Greystones on, however, only [illegible] days.

AUGUST.—August, 1889, proved a wet, cold, and stormy month. Rain fell on as many as 22 days—the fall amounting to 1·[illegible] inches on the 10th, and to 1·[illegible] inches on the 19th. The mean temperature was 1·[illegible]° below the average, and strong or squally winds between S.W. and N.W. prevailed on no fewer than 14 days. As regards temperature conditions were almost identical with those observed in August, 1888.

In Dublin the mean temperature (5[illegible]°) was decidedly below the average ([illegible]°); the mean dry bulb readings at 9 a.m. and 9 p.m. were 57·[illegible]°. In the twenty-four years ending with 1888, August was coldest in 1881 (M. T. = 57·0°), and warmest in 1871 (M. T. = 62·0°). In 18[illegible], the M. T. was only 57·1°; in 1879 (the "cold year"), it was 57·7°; in 1887, it was 60·[illegible]°; and in 1888, it was 58·2°.

The mean height of the barometer was 29·[illegible] inches, or 0·0[illegible] inch below the average value for August—namely, 29·90[illegible] inches. The mercury marked 30·[illegible] inches at 9 a.m. of the 31st, and fell to 29·1[illegible] inches at 9 p.m. of the 21st. The observed range of atmospherical pressure was therefore 1·1[illegible] inches—that is, a little more than an inch and one-tenth. The mean temperature deduced from daily readings of the dry bulb thermometer at 9 a.m. and 9 p.m. was 57·[illegible]°, or exactly the same as in August, 1888, and 0·[illegible]° below the value for July, 1889; that calculated by Kaemtz's formula from the means of the daily maxima and minima was 57·[illegible]°, or 1·[illegible]° below the average mean temperature for August, calculated in the same way, in the twenty years 1865–84, inclusive (59·[illegible]°). The arithmetical mean of the maximal and minimal readings was [illegible], compared with a twenty-three years' average of 60·[illegible]°. On the 1st the thermometer in the screen rose to 71·[illegible]°—wind, S.W.; on the 24th the temperature fell to 48·[illegible]°—wind, N.N.W. The minimum on the grass was 4[illegible]° on the latter date. The rainfall was 6·7[illegible] inches, distributed over 23 days. The average rainfall for August in the twenty-three years, 1865–87, inclusive, was 2·7[illegible] inches, and the average number of rainy days was 17·[illegible]. The rainfall, therefore, was more than twice the average, and the rainy days were also considerably in excess of the average. In 1874 the rainfall in August was very large—4·[illegible] inches on 22 days—and in 1868 also 4·745 inches fell on, however, only 15 days. On the other hand, in 1884, only ·777 inch was measured on 8 days. In 1887, 1·[illegible] inches of rain fell on 16 days, and in 1888, 1·270 inches on 15 days.

High winds were noted on so many as 14 days, and attained the force of a gale on [illegible] two occasions—namely, the 4th and 26th. The atmosphere was foggy on the 31st. Thunder was heard on the 7th and 11th, and lightning was seen on the 26th. Temperature reached or exceeded 70° in the screen on only two days—or 3 fewer than in August, 1887, and [illegible] fewer than in August, 1888.

On Thursday, the 1st, heavy showers began to fall **at intervals, and continued to** do so during the two next days.

Very unsettled showery weather prevailed throughout the week ending Saturday, the 10th, and from Monday evening until Wednesday morning the wind was fresh or strong in force. Rain fell in drenching showers, and occasional thunder and lightning accompanied the showers. The barometer stood relatively high in France and Germany, while a series of atmospherical depressions of some intensity passed in an easterly, or north-easterly, direction across the British Isles, the North Sea, and Scandinavia. Hence the prevalence of squally southwesterly (S. to W. and N.W.) winds and heavy thunder showers. In Dublin Sunday and Friday were tolerably fine days. Heavy showers fell on Monday and Tuesday. There was a moderate gale from W. to N.W. on Monday night. At 4 p.m. of Wednesday a local easterly sea breeze caused dense raincloud to form and at 5 30 p.m. a flash of lightning and clap of thunder were followed by torrents of rain—the measurement from the shower being ·430 of an inch in less than half an hour. On Saturday rain again fell in large quantities over the city for several hours. In Dublin the mean height of the barometer was 29·[illegible] inches, pressure ranging from 30·077 inches at 9 p.m. of Wednesday (wind, W.), to 29·[illegible] inches, at 9 p.m. of Saturday (wind, W.). The mean dry bulb temperature at 9 a.m. and 9 p.m. was 57·[illegible]°. The mean of the daily highest and lowest shade temperature was 5[illegible]°. The thermometer rose to 6[illegible]° in the shade on Wednesday, and fell to 5[illegible]° on Thursday. Rain fell on six days to the large amount of 2·[illegible] inches—the maximal fall in 24 hours being 1·0[illegible] inches on Saturday (wind, S.E.). Thunder and lightning occurred on Wednesday.

A cyclonic distribution of atmospherical pressure and changeable weather held during the second week (11th–17th inclusive). On Sunday, the 11th, a complex area of low pressure lay over England and the North Sea—in Ireland, northerly winds and sharp thunderstorms were felt and the weather was everywhere unsettled. By Tuesday morning the centre of this system, which had become much deeper, had reached the Baltic, the barometer reading only 29·[illegible] inches at Wisby, in Gothland. The weather was now fine in Ireland, but in the course of Tuesday night a new disturbance arrived off Donegal, accompanied by a heavy downpour of rain. Fresh or squally S.W. to N.W. winds prevailed until Thursday evening. On Friday afternoon another depression approached Ireland from the Atlantic, causing heavy showers and a renewal of squalls—first from S.W., and then from W. In Dublin the mean height of the barometer during the week was 29·[illegible] inches, and pressure ranged from 30·[illegible] inches, at 9 a.m. of Tuesday (wind, N.N.W.), to 29·[illegible] inches, at 9 p.m. of Friday (wind, S.W.). The mean dry bulb temperature at 9 a.m. and 9 p.m. was 5[illegible]°. The mean of the daily highest and lowest readings of the screened thermometers was 5[illegible]°. The highest shade temperature was 6[illegible]° on Friday; the lowest was 5[illegible]° on Tuesday. Rain fell on four days to the amount of ·5[illegible] inch—the heaviest fall in 24 hours being ·[illegible] inch on Tuesday.

Exceedingly broken and unseasonable weather held throughout the third week (18th–24th inclusive). Large quantities of rain fell in all parts of the United Kingdom, France, and Scandinavia, and westerly (S.W. to N.W.) gales of unusual severity for August were reported from the coast stations. On Sunday, the 18th, the isobars ran in almost straight lines from W. to E. across the British Isles, the North Sea, Denmark, and Northern Germany—the resulting westerly wind was high, but no heavy rainfall occurred. During the remainder of the week, however, a series of deep atmospherical depressions passed northeastwards across Ireland and Scotland, and to these the gales and rains were due. By Tuesday morning about 2 inches of rain had fallen over the greater part of Ireland, and a similar downpour had begun in Scotland. On Wednesday another depression travelled quickly across the North of Ireland and of England, renewing the bad weather. The remainder of the week was cold, showery, and squally, with strong N.W. winds. In Dublin the mean height of the barometer was 29·650 inches, the readings varying between 29·[illegible] inches at 9 a.m. of Friday (wind, W.N.W.), and 29·1[illegible] inches at 9 p.m. of Wednesday (wind, W.). The mean temperature at 9 a.m. and 9 p.m. was 5[illegible]°. The mean of the daily extreme temperatures was 5[illegible]°. The shade thermometers rose to 6[illegible]° on Sunday and Tuesday, and fell to 4[illegible]° on Friday. Rain fell every day, the total measurement being 2·574 inches, of which 1·[illegible] inch was registered on Monday.

A marked improvement in the weather took place in the week (25th–31st inclusive), at all events over the greater part of the country. In the northwest and north of Ireland and in Scotland, indeed, bad weather prevailed from Tuesday to Friday, owing to the passage of a [illegible] eastwards across these districts. But in the east and south of Ireland and in England very little rain fell, and a return of summer temperatures was observed. On Friday the thermometer rose to 69° in London and to 68° at Oxford and Cambridge. The wind was N.W. on Sunday and Monday. It then backed to W. and S.W., and freshened considerably, under the influence of a depression which skirted the west coasts of Ireland and Scotland, causing gales and rains in exposed places. On Friday the wind died down, and a light easterly breeze and clear sky succeeded a calm foggy morning on Saturday. In Dublin the mean height of the barometer was 30·[illegible] inches, pressure ranging from 29·[illegible] inches at 9 a.m. of Sunday (wind, N.N.W.), to 30·[illegible] inches at 9 a.m. of Saturday (wind, calm). The mean dry bulb temperature at 9 a.m. and 9 p.m. was 57·[illegible]°. The arithmetical mean of the daily maximal and minimal temperatures in the screen was 57·[illegible]°. The shade thermometer rose to 6[illegible]° on Wednesday, having fallen to 4[illegible]° on Sunday. Rain fell on two days to the amount of only ·020 inch—·010 inch on Monday, and ·010 inch on Friday. There was a smoke fog in the city on Saturday morning, and sheet lightning was seen on Friday night.

The rainfall in Dublin during the eight months ending August 31st has amounted to 18·[illegible] inches on 1[illegible] days, compared with [illegible] inches on [illegible] days during the same period in 1887, 1[illegible] inches on 121 days in 1888, and a twenty-three years' average of 17·815 inches, on 12[illegible] days.

At Greystones, Co. Wicklow, the rainfall in August, 1889, was [illegible] inches, distributed over 17 days. Of this quantity, [illegible] inches fell on the 1[illegible]th, ·[illegible] inch on the 2[illegible]st, and ·[illegible] inch on the 2nd. Since January 1, 20·000 inches of rain have fallen at Greystones, on, however, only [illegible] days.

At Carrickmines, County Dublin, the rainfall in August was [illegible] inches, on [illegible] days; and since January 1, 20·[illegible] inches have fallen at that station on 121 days.

September.—As in 1888, so in 1889, September proved a fine month, with a high mean barometer, winds from polar quarters—chiefly N.W., a moderate rainfall, and normal temperature. At the beginning of the month fogs were prevalent along the sea coast, and towards the close blowing and squally winds were experienced on several days.

In Dublin the mean temperature (5[illegible]°) was exactly equal to the average (5[illegible]°); the mean dry bulb readings at 9 a.m. and 9 p.m. were 55·[illegible]°. In the twenty-four years ending with 1888, September was coldest in 1866 and in 1882 (M. T. = 5[illegible]°), and warmest in 1865 (M. T. = 61·[illegible]°). In 1886 the M. T. was as high as 58·[illegible]°; in the year 1879 (the "cold year"), it was 54·8°; in 1887, it was 54·0°; and in 1888, it was 54·[illegible]°.

The mean height of the barometer was 30·064 inches, or 0·147 inch above the average value for September—namely, 29·917 inches. The mercury rose to 30·455 inches at 9 a.m. of the 1[illegible]th, and fell to 29·504 inches at 9 p.m. of the 23rd. The observed range of atmospherical pressure was, therefore, ·951 of an inch—that is, a little less than one inch. The mean temperature deduced from daily readings of the dry bulb thermometer at 9 a.m. and 9 p.m. was 55·[illegible]°, or 2·[illegible]° below the value for August, 1889; that calculated by Kaemtz's formula from the means of the daily maxima and minima was 5[illegible]·8°, or 0·[illegible]° below the average mean temperature for September, calculated in the same way, in the twenty years, 1865–84, inclusive (55·[illegible]°). The arithmetical mean of the maximal and minimal readings was 5[illegible]·[illegible]°, compared with a twenty-three years' average of 5[illegible]·[illegible]°. On the 11th the thermometer in the screen rose to 69·7°—wind, W.S.W.; on the 2[illegible]th the temperature fell to [illegible]°—wind, N.W. The minimum on the grass was [illegible]° on the same date. The rainfall was only 1·[illegible] inches, distributed over 1[illegible] days—both rainfall and rainy days were much below the average. The average rainfall for September in the twenty-three years, 1865–87, inclusive, was [illegible] inches, and the average number of rainy days was 15·[illegible]. In 1871 the rainfall in September was very large—[illegible] inches on, however, only 1[illegible] days. On the other hand, in 18[illegible], only ·[illegible] inch was measured on but three days.

High winds were noted on [illegible] days, but attained the force of a gale on only one occasion, the 1[illegible]th. The atmosphere was foggy on the 1st, 2nd, 4th, 5th, [illegible]th, and 1[illegible]th. Heavy dews and wet fogs were experienced at night about the beginning of the month.

Quiet, but chiefly dull or foggy, weather prevailed the week ending Saturday, the 7th. At the beginning and close the distribution of atmospherical pressure in Western Europe was anticyclonic, with resulting light variable winds or calms. In the interval between Monday and Thursday, however, a number of shallow, complex depressions passed slowly eastwards across the United Kingdom, causing close thundery weather, with much fog and gloom, and a damp air. Dublin escaped all electrical disturbances, but in the S.W. of Ireland and over the greater part of England thunderstorms were felt—the rainfall accompanying them not being specially heavy, except at a few stations such as Cambridge, where ·[illegible] inch of rain was measured on Tuesday morning. Near Dublin damp vapour-fogs hung about the coast from Wednesday to Saturday—this latter day proving brilliant in the extreme. In Dublin the mean height of the barometer was 30·[illegible] inches, pressure ranging from 29·[illegible] inches at 9 p.m. of Monday (wind S.E.) to 30·[illegible] inches at 9 a.m. of Friday (wind S.E. by E.). The mean dry bulb temperature at 9 a.m. and 9 p.m. was 5[illegible]°, while the arithmetical mean of the daily maxima and minima was 5[illegible]·[illegible]°. Temperature in the shade was [illegible] on Wednesday (6[illegible]·[illegible]°), and lowest on Sunday (4[illegible]·[illegible]°). Rain fell in measurable quantity on Monday to the amount of ·020 inch.

There was but little change in the weather of the second week, 8th to 14th, inclusive. For the most part fine bright days prevailed, with light to moderate S.W. to Northerly breezes. On Wednesday it was beautifully fine and summerlike. Temperature rose to 66·7°, and is reported to have risen to 75° at some of the Irish stations. On Saturday the barometer rose slowly with clouded sky, and Northerly to North Easterly wind; some slight drizzling rain fell; but although the cloud canopy held persistently during the day the afternoon was fine. In Dublin the mean height of the barometer was 30·[illegible] inches, pressure ranging from 29·[illegible] inches at 9 p.m. of Sunday (wind S.W.) to 30·[illegible] inches at 9 a.m. of Saturday (wind N.). The mean dry bulb temperature at 9 a.m. and 9 p.m. was 5[illegible]·[illegible]°. The arithmetical mean of the daily highest and lowest temperatures in the shade was 30·8°. On Wednesday the thermometer in the screen rose to 69·F. On Monday the lowest reading for the week occurred and was 38·4°. Rain was measured on four days, the total amount being, however, only ·139 inch, of which ·108 inch was entered to Sunday, the 8th.

There was a marked contrast in the weather of the third week (15 to 21st inclusive) to that of the preceding one. Monday, the 9th, was cloudy but fine. Tuesday was bright and fine, but much cirrus cloud was visible, foretelling the approach of a large depression, which subsequently arrived. Towards evening on Wednesday, the 18th, the barometer fell quickly, the sky clouded over and the wind veering to the north, freshened to a moderate gale. Heavy rain fell during the night and part of Thursday. For the rest of the week the weather remained broken, being for the most part cloudy and cold; heavy showers at intervals; wind varying from N.W. to N.E., and squally. Temperature fell considerably, and continued low. Slight ground frost was experienced in some parts of England at night. In Dublin the mean height of the barometer during the week was 30·011 inches, pressure decreasing from 30·155 inches on Sunday at 9 a.m. (wind E.) to 29·841 inches on Saturday at 9 p.m. (wind N.). The mean dry bulb temperature at 9 a.m. and 9 p.m. was 51·1°, the mean of the daily maxima and minima being 53·7°. The shade temperature rose to 62·3° on Wednesday and fell to 40·3° on Saturday. The rainfall amounted to ·413 inch, on four days. The largest daily rainfall was ·288 inch on Wednesday.

The weather of the period, from Sunday, the 22nd, to Saturday, the 28th inclusive, was in all respects changeable, and temperature in particular was unsteady. At the beginning the weather was clear and cold, with light northerly winds. On Monday a depression appeared off the S.W. of Ireland, whence it travelled across the S. of England to the North Sea and Denmark. It caused a considerable rainfall in Dublin (·336 inch), where the wind shifted to the northward in its wake. Over the S. and E. of England however the downpour was much greater—1·9 inches being registered at Cambridge in the 24 hours ending 8 a.m. of Wednesday. Another brief spell of clear, cold weather was followed on Thursday by freshening westerly (S.W. to N.W.) winds, a sudden and marked rise of temperature, and clouds and rain in Ireland and Scotland. On this occasion the precipitation was greatest in the N.W. and N. of Ireland—1·0 inch of rain being registered at Malin Head at 8 a.m. of Friday. Towards evening of this day the wind veered to N.W., temperature gave way, and the weather improved. In Dublin the mean height of the barometer was 29·896 inches—the extremes being—highest, 30·156 inches at 9 p.m. of Wednesday (wind, W.N.W.); lowest, 29·464 inches at 9 p.m. of Monday (wind, E.S.E.). The mean dry bulb temperature at 9 a.m. and 9 p.m. was 50·6°. The mean of the daily maxima and minima was 50·0°. The shade thermometer rose to 59·6° on Thursday, having fallen to 38·9 on Wednesday, when the grass minimum was 32·9°. Rain fell on four days to the amount of ·471 inch, the maximal fall in 24 hours being ·306 inch on Monday.

The last two days of the month were dry and bracing, with strong northerly winds, and alternate cloud and sunshine.

The rainfall in Dublin during the nine months ended September 30th has amounted to 19·084 inches on 147 days, compared with 10·906 inches on 113 days during the same period in 1887, 17·092 inches on 131 days in 1886, and a twenty-three years' average of 19·861 inches, on 148·1 days.

At Greystones, Co. Wicklow, the rainfall in September, 1889, was 1·030 inches, distributed over only 7 days. Of this quantity, ·760 inch fell on the 23rd. Since January 1, 21·120 inches of rain have fallen at Greystones, on, however, only 90 days.

OCTOBER.—A cold, rainy, squally month—westerly and south-westerly winds prevailing at first while easterly and north-easterly winds held from the 10th to the 18th. A gale of exceptional violence blew over Ireland on the night of the 6th, the wind rising from S.S.E. and subsequently veering to W. and W.N.W. There was another storm—less destructive in its effects—on the night of the 31st.

In Dublin the mean temperature (48·1°) was considerably below the average (49·8°); the mean dry bulb readings at 9 a.m. and 9 p.m. were 47·8°. In the twenty-four years ending with 1888, October was coldest in 1880 (M. T. = 45·9°), and in 1885 (M. T. = 45·6°), and warmest in 1876 (M. T. = 53·1°). In 1886, the M. T. was as high as 52·0°; in the year 1879 (the "cold year") it was 49·7°. In 1887 it was as low as 47·2°; and in 1888, it was 49·3°.

The mean height of the barometer was 29·873 inches, or 0·174 inch below the average value for October—namely, 29·347 inches. The mercury rose to 30·293 inches at 9 p.m. of the 15th, and fell to 28·780 inches at 4 a.m. of the 7th. The observed range of atmospherical pressure was, therefore, 1·50 inches—that is, a little more than an inch and one half. The mean temperature deduced from daily readings of the dry bulb thermometer at 9 a.m. and 9 p.m. was 47·5°, or 7·6° below the value for September, and 10·3° below that for August, 1889; that calculated by Kaemtz's formula from the means of the daily maxima and minima was 47·2°, or 2·0° below the average mean temperature for October, calculated in the same way, in the twenty years, 1865–84, inclusive (49·2°). The arithmetical mean of the maximal and minimal readings was 48·1°, compared with a twenty-three years' average of 49·8°. On the 7th the thermometer in the screen rose to 67·4°—wind, S.S.W.; on the 14th the temperature fell to 33·8°—wind calm. The minimum on the grass was 28·6° on the 14th. The rainfall was 4·552 inches, distributed over 22 days—both rainfall and rainy days were largely in excess. The average rainfall for October in the twenty-three years, 1865–87, inclusive, was 3·111 inches, and the average number of rainy days was 17·4. In 1880 the rainfall in October was very large—7·358 inches on 18 days. In 1875, also, 7·049 inches fell on 26 days. On the other hand, in 1884, only ·515 inch was measured on but 14 days, and in 1869 only ·866 inch on 15 days. In 1888 the rainfall was 1·237 inches on 16 days. A lunar rainbow was seen on the 6th. High winds were noted on 18 days, and attained the force of a gale on four occasions, the 6th, 7th, 8th, and 31st. The atmosphere was more or less foggy in Dublin on the 10th, 14th, and 25th. Lightning was seen on the evenings of the 6th and 31st. Thunder was heard about 8 p.m. of the 31st. Hail fell with heavy showers of rain on the 7th and 28th.

Tuesday, the 1st, was a cool, fine day, with light northerly winds and dry air. On Wednesday, the 2nd, a complex series of atmospherical depressions came in from the westward over the United Kingdom, and in consequence the weather fell into a very dull and rainy state—more particularly as areas of high pressure existed both in the extreme North of Europe and over the Peninsula and South of France. At 8 a.m. of Friday, the 4th, the distribution of temperature was peculiar—at Loughborough in Leicestershire the thermometer marked 34°, at Yarmouth it was 41°, while at Bodö in Norway (within the Arctic circle) it was as high as 58°. The rainfall was very heavy at many stations, and thunder and lightning were reported from the East and South of England on Thursday night and on Friday.

At first unsettled, stormy, and rough, the weather afterwards became quiet, cool, and fair in the week ending Saturday, the 12th. On Sunday, the 6th, the sky and clouds wore a very threatening look, and in the afternoon the barometer fell at the rate of a tenth of an inch an hour, as a very deep depression approached the North of Ireland from the Atlantic. In Dublin rain fell steadily from 8 p.m., and at 10 p.m. a furious gale burst upon the city. The wind rose from S.S.E. to S., but veered to S.W., and finally to W.N.W., as the night wore on. At 10 30 p.m. a lunar rainbow was seen. The gusts were at times of hurricane force, so that much damage was done, especially to trees which were still in full leaf. The minimal pressure was 28·31 inches at Belmullet, Co. Sligo; in Dublin the barometer read 28·79 inches at 8 a.m. of Monday. Hail fell in the afternoon of this day, which was rough and squally. During Tuesday fine weather prevailed, but lightning was seen after nightfall. A new and deep depression arrived off the North of Ireland on Tuesday, causing a renewal of bad weather. After this the barometer rose slowly in Ireland, while some secondary disturbances kept the weather unsettled in England. At 8 a.m. of Friday the thermometer read 31° at Loughborough, 35° at Parsonstown, 43° in London (where there was a dense fog), and 48° at Prawle Point. In Dublin the mean height of the barometer was only 29·[illegible] inches—pressure decreased from 29·787 inches at 9 a.m. of Sunday (wind, W.) to 28·790 inches about 4 a.m. of Monday (wind, S.W., force 11). The mean dry bulb temperature at 9 a.m. and 9 p.m. was 48·9°. The mean of the daily highest and lowest temperatures was 48·6°. The thermometer in the screen rose to 57·4° on Monday and fell to 38·1° on Friday. Rain fell on four days to the amount of ·795 inch, of which ·494 inch was referred to Sunday.

The earlier part of the week ending Saturday, the 19th, was calm and fair in Dublin—during the latter part a rainstorm of exceptional violence prevailed. On Monday morning the first hoar-frost of the season was observed, the thermometer falling to 28·4° on the grass and to 35·2° in the screen. On Tuesday an extensive, but not deep, depression lay off the northwest of Ireland, so that the weather fell into an unsettled state, with S.W. winds and showers. Wednesday proved fine and bright, after a dull, damp, morning. In England, a heavy rainfall occurred on this day. Thursday was at first fine, afterwards showery. At night the barometer gave way quickly, as a deep V-shaped depression slowly approached Ireland from the westward. Squally S.E. winds sprang up and a continuous rain set in, which in Dublin lasted to the close of the week. At noon on Saturday, the wind shifted from E. by S. to W. by S., as the centre of the low pressure system passed over the city. In Dublin the mean height of the barometer during the week was 29·[illegible] inches. Pressure ranged between 29·[illegible] inches at 9 p.m. of Sunday (wind, W.N.W.), and 28·[illegible] inches at 9 p.m. of Saturday (wind, W.). The mean dry bulb temperature at 9 a.m. and 9 p.m. was 48·[illegible]°. The mean of the daily highest and lowest readings of the thermometer was [illegible]°. Temperature in the screen fell to 34·2° on Monday and rose to 57·9° on Tuesday. Rain fell on five days to the large amount of 2·013 inches—the measurement on Friday was ·730 inch, and that on Saturday was [illegible] inch.

The week ended the 26th was again a period of unsettled rainy weather—especially on the east and south coasts of Great Britain and Ireland. In the central districts of the latter country, it was much finer and towards the close of the week some sharp frosts were experienced—the shade thermometer falling at Parsonstown, King's County, to 30° on Thursday and 29° on Friday. The general distribution of atmospherical pressure during the greater part of the week was peculiar—two areas of high pressure existed over the extreme North of Europe and over the Mediterranean and Algeria, while a band of low pressure—in which there were several distinct minima—stretched across Central Europe from the south of the United Kingdom to Southern Russia. It was in this trough of low barometer that the weather was so bad. Towards the end of the week a depression was found over the Peninsula. Winter may be said to have set in over Northern Russia on Tuesday, the 22nd, when the thermometer marked 28° at 8 a.m. at Archangel. In Dublin every day was more or less showery, and on Monday evening vivid lightning was frequently seen to the southward, accompanied by thunder at 8 p.m. On Friday night there were heavy hail showers. The mean barometer reading was 29·886 inches, pressure ranging from 29·[illegible] inches at 9 a.m. of Sunday (wind, E. by N.) to 30·283 inches at 9 p.m. of Friday (wind, N.E. by E.). The mean dry bulb temperature at 9 a.m. and 9 p.m. was 47·1°. The mean temperature deduced from the daily extremes was 47·1°. The thermometer rose to 58·0° on Monday, and fell to 38·1° on Thursday. Rain was measured on every day, the total quantity being ·[illegible] inch, of which ·2[illegible] inch was registered on Friday, and ·324 inch on Monday.

The weather of the last few days of the month (27th–31st) varied considerably. A northerly air current held on Sunday and Monday, the 27th and 28th, in connection with a cyclonic system over the English Channel and Brittany. It caused clouds and showers on the east coasts of Great

Britain and Ireland, but inland clear skies and frosty nights prevailed. On the 29th, S.W. winds spread over the whole country, temperature rose fast, and showers and squalls set in. A severe storm occurred on the night of the 31st.

The rainfall in Dublin during the ten months ending October 31st has amounted to 24·793 inches on 166 days, compared with 12·859 inches on 128 days during the same period in 1887, 19·219 inches on 167 days in 1888, and a twenty-three years' average of 22·914 inches on 162 days.

At Greystones, Co. Wicklow, the rainfall in October, 1889, was no less than 6·035 inches, distributed over 22 days. Of this quantity, 1·140 inches fell on the 8th, 1·250 inch on the 18th, and ·870 inch on the 19th. Since January 1st, 25·045 inches of rain have fallen at Greystones, on, however, only 112 days.

NOVEMBER.—No greater contrast can be imagined than the weather of November, 1889, compared with that of November, 1888, which was the wettest and most stormy November experienced in Dublin for more than a quarter of a century. The past month was, on the contrary, calm, dull, and comparatively mild and rainless. An anticyclone held over western Europe with unusual persistence, and the warm, damp, southerly and westerly winds of the system brought a second summer to the British Islands after the cold and rain of October. Only at the close of the month was a spell of wintry weather experienced in the rear of a remarkable series of V-shaped depressions which passed eastwards across north-western Europe.

In Dublin the mean temperature (46·6°) was decidedly above the average (44·5°); the mean dry bulb readings at 9 a.m. and 9 p.m. were 45·7°. In the twenty-four years ending with 1888, November was coldest in 1878 (M. T. = 38·8°), and in 1879 (M. T. = 42·4°), and warmest in 1881 (M. T. = 50·2°). In 1866, the M. T. was as high as 49·4°; in the year 1879 (the "cold year") it was 40·9°; in 1887 it was as low as 43·0°; and in 1888, it was as high as 47·8°.

The mean height of the barometer was 30·131 inches, or ·234 inch above the average value for November—namely, 29·897 inches, and ·501 inch above the mean pressure in November, 1888—viz., 29·630 inches. The mercury rose to 30·612 inches at 9 p.m. of the 15th, and fell to 28·830 inches at 7 p.m. of the 24th. The observed range of atmospherical pressure was, therefore, 1·782 inches—that is, a little more than an inch and three-fourths. The mean temperature deduced from daily readings of the dry bulb thermometer at 9 a.m. and 9 p.m. was 45·7°, or only 1·5° below the value for October, and 9·4° below that for September, 1889; that calculated by Kaemtz's formula from the means of the daily maxima and minima was 46·6°, or 1·9° above the average mean temperature for November, calculated in the same way, in the twenty years, 1865–84, inclusive (44·7°). The arithmetical mean of the maximal and minimal readings was 46·8°, compared with a twenty-three years' average of 44·8°. On the 7th the thermometer in the screen rose to 58·7°—wind, W.S.W.; on the 28th the temperature fell to 29·0°—wind, W. The minimum on the grass was 23·9° on the 28th. The rainfall was only ·319 inch, distributed over 9 days—both rainfall and rainy days fell largely short of the values for the preceding month of October—viz., 4·552 inches on 22 days. The average rainfall for November in the twenty-three years, 1865–87, inclusive, was 2·440 inches, and the average number of rainy days was 16·9. The rainfall and the rainy days, therefore, were very much below the average. In 1878 the rainfall in November was large—5·318 inches on 20 days. In 1872, also, 3·424 inches fell on 24 days; in 1867, 3·013 inches fell on 18 days, and in 1888, 6·540 inches fell on 26 days. On the other hand, in 1870, only 1·215 inches were measured on but 11 days, and in 1879 only 1·231 inches on but 10 days. The vast excess of the rainfall in 1888 and the equally marked deficit in 1889 are evident from these figures.

A lunar halo appeared on the 4th, and a solar halo on the 14th. High winds were noted on 3 days, but attained the force of a gale on only one occasion, the 1st. The atmosphere was more or less foggy in Dublin on the 11th, 13th, 14th, 15th, 17th, 18th, 19th, 23rd, and 30th. Lightning was seen on the night of the 25th. Hail fell on the 1st. Sleet and snow fell on the 26th and 27th. An aurora borealis was seen on the night of the 26th.

The month opening with a strong S.W. gale—the only storm which was felt throughout in Dublin. It was followed by a temporary fall of temperature and some cold showers.

Westerly winds and dull, mild weather were the leading features of the week ending Saturday, the 9th. At first an irregular depression was passing across Great Britain eastwards; it was rainy in England, showery in Ireland, fair and cold over the greater part of Scotland. Monday, the 4th, was cool and fair in Dublin. Some rain fell on Tuesday morning, but this day also proved fine. On Wednesday morning a large Atlantic anticyclone was found off the South of Ireland, whence it extended across the Bay of Biscay to the North of Spain. On the afternoon of this day temperature rose decidedly, and the weather became warm and damp—remaining so until the end of the week. At 8 a.m. of Thursday the thermometer read 37° at Paris, but 55° at Aberdeen, where it rose to 62° later in the day. This remarkable reading for November was also recorded at Loughborough. Even in Northern Russia very little frost was felt during this period. In Dublin the mean height of the barometer was 30·178 inches. Pressure increased from 29·544 inches, at 9 a.m. of Sunday (wind, W. by S.) to exactly one inch higher, namely, 30·544 inches at 9 a.m. of Friday (wind, W.S.W.). The mean dry bulb temperature at 9 a.m. and 9 p.m. was 49·7°, and the mean of the daily extremes of temperature was 50·6°. The thermometer in the shade rose to 58·7° on Thursday, having fallen to 37·9° on Monday. Rain fell on two days to the amount of ·098 inch, of which ·090 inch was measured on Friday. At 8 a.m. of that day the barometer read 30·61 inches at Lorient, and 30·60 inches at Rochefort and at Valentia, in Kerry.

The second week (10th–16th, inclusive) was a period of singularly mild, quiet, weather. The distribution of atmospherical pressure was anticyclonic, except in the far North, where depressions were found—of, however, no great intensity. At the beginning of the period, the centre of a large anticyclone, in which the barometer was as high as 30·50 inches, lay off the extreme S.W. of Ireland.

The weather was dull and mild. This high pressure system then moved eastwards, decreasing at the same time in intensity. By Tuesday morning its centre had reached Prussia. A sharp morning frost with fog prevailed over central England on this day, but in Ireland it remained very mild, dull, and damp. Up to and including Friday, the 15th, the anticyclone held over the continent. On this day a V-shaped depression of no great depth came in over Ireland from the Atlantic. A squally S.W. wind and showers were the result, but the barometer quickly recovered, and Saturday was bright and cold, with light winds from N.W. to N.E., and a smoke fog at night. In Dublin the mean height of the barometer was 30·351 inches. Pressure ranged from [illegible] inches at 3 p.m. of Thursday (wind, S.S.W.), to 30·615 inches at 9 p.m. of Saturday (wind, calm). The mean dry bulb temperature at 9 a.m. and 9 p.m. was 45·9°. The mean of the daily highest and lowest shade temperatures was 46·1°. The thermometer rose to 53·7° on Friday, and fell to 41·0° on Saturday. Rain fell on Friday night to the amount of ·036 inch—this was the only measurable rainfall during the week. The air was foggy on Monday, Wednesday, Thursday, and Saturday. A solar halo was seen on Thursday forenoon.

Quiet, mild, damp, and foggy weather held during the greater part of the third week (17th–23rd inclusive). Only on the evening of Thursday, the 21st, did the wind freshen to a steady or stiff breeze, which reached the force of a gale in the West of Ireland. In England, Germany, and France temperature decreased greatly as the week advanced; but the season remained singularly open in Scotland and even in Scandinavia. This was brought about by the persistent presence of a large anticyclone over Western Europe. In its centre, which lay over Germany, the barometer remained above 30·70 inches from Sunday night until Thursday afternoon—the highest reading of all being 30·75 inches, recorded at Berlin and Wiesbaden on Wednesday. On Thursday night a V-shaped depression advanced over Ireland from the S.W. It caused the strong winds above referred to and a heavy fall of rain on Friday. In the rear of this low pressure system, of no great depth, the barometer again rose quickly and the sky cleared with a sudden fall of temperature. Saturday was at first fair and bright, but in the afternoon a dense vapour fog formed. In Dublin the mean height of the barometer was 30·413 inches, pressure ranging from 30·610 inches at 9 a.m. of Sunday (wind, calm) to 30·035 inches at 9 a.m. of Friday (wind, S. by W.). The mean dry bulb temperature at 9 a.m. and 9 p.m. was 46·3°. The mean of the daily highest and lowest temperatures was 47·0°. The screened thermometers rose to 57·5° on Friday, having fallen to 36·4° on Sunday. Rain was measured on three days, the total quantity being ·406 inch, and the maximal fall in 24 hours being ·390 inch on Friday.

A remarkable and sudden change from autumn to winter took place at the beginning of the last week ending Saturday, the 30th. It proved wintry in the extreme. On Sunday a general and decided fall of the barometer over Western Europe ushered in a complex series of V-shaped depressions, which at 8 a.m. of Monday extended as a trough of low pressure from the W. coast of Norway to the S. of France. In front of this trough the wind was strong from S. and S.W., with comparatively high temperature and rain—in its rear, fresh N.W. and N. winds were blowing with a keen, dry air and showers of sleet, hail and snow. On Monday night thunder and lightning occurred in the North-west of England. At 8 a.m. of Tuesday the barometer was down to 28·53 inches at Bodö in Norway—a gale prevailing, with snow squalls in many places. At night discharges of thunder and lightning were reported from Ireland and Scotland. Wednesday was a very cold day—temperature did not rise above 37·8° in Dublin. On Thursday an anticyclone passed south-eastwards across Munster, temperature rose and S.W. winds became established in Ireland. In Dublin the mean height of the barometer was 29·864 inches—pressure ranging between 29·290 inches at 7 p.m. of Sunday (wind, S.W.) and 30·257 inches at 9 a.m. of Thursday (wind, W.). The mean dry bulb temperature at 9 a.m. and 9 p.m. was 38·9°. The mean of the daily highest and lowest thermometer readings was 39·7°. Temperature rose to 52·5° on Sunday, and fell to 29·4° on Thursday. Lightning was seen on Monday night; rain fell on Sunday to the amount of ·276 inch. Light sleet and snow showers on Monday, Tuesday, and Wednesday yielded no measurable rainfall.

The rainfall in Dublin during the eleven months ending November 30th has amounted to 22·718 inches on 174 days, compared with 18·378 inches on 151 days during the same period in 1897, 29·768 inches on 193 days in 1898, and a twenty-three years' average of 25·224 inches on 177·4 days.

At Greystones, Co. Wicklow, the rainfall in November, 1899, was 1·380 inches distributed over 9 days. Of this quantity ·440 of an inch fell on the 23rd, and ·340 of an inch on the 26th. Since January 1, 22·135 inches of rain have fallen at Greystones on, however, only 121 days.

December.—A quiet, mild, but damp month, with prevalent south-westerly winds (that is, winds from points between S. and W.). These conditions were determined by the persistence of an anticyclone to the eastward and south-eastward, while successive systems of low pressure skirted the western shores of Ireland, Scotland, and Norway in their passage northeastwards across the Atlantic Ocean and the Norwegian Sea. In central and south-eastern England sharp frost was felt towards the close of the month, while the weather remained open in Ireland and Scotland.

In Dublin the mean temperature (42·8°) was—as in November—above the average (41·1°); the mean dry bulb readings at 9 a.m. and 9 p.m. were 42·3°. In the thirty-five years ending with 1899, December was coldest in 1878 (M. T. = 32·8°), and in 1874 (M. T. = 36·8°), and warmest in 1865 (M. T. = 46·2°). In 1886, the M. T. was as low as 37·7°; in the year 1879 (the "cold year") it was also 37·6°. In 1897 the M. T. was 39·9°; in 1898 it was 43·6°, or very nearly the same as in 1899.

The mean height of the barometer was 30·096 inches, or 0·214 inch above the average value for December—namely, 29·882 inches. The mercury rose to 30·730 inches at 9 p.m. of the 5th, and fell to 29·260 inches at 9 p.m. of the 16th. The observed range of atmospherical pressure was, therefore, 1·470 inches—that is, a little less than one inch and a-half. The mean temperature deduced from daily readings of the dry bulb thermometer at 9 a.m. and 9 p.m. was 43·2°, or 2·4° below the value for November, and 4·9° below that for October, 1889; that calculated by Kaemtz's formula from the means of the daily maxima and minima was 43·0°, or 2·4° above the average mean temperature for December, calculated in the same way, in the twenty years, 1865–84 inclusive (40·6°). The arithmetical mean of the maximal and minimal readings was 43·8°, compared with a twenty-three years' average of 41·1°. On the 17th the thermometer in the screen rose to 56·0°—wind, W.S.W.; on the 6th and the 12th the temperature fell to 30·6°—wind S.S.W. on both occasions. The minimum on the grass was 22·9° on the 12th. There were 4 days of frost in the screen and 20 days of frost on the grass.

The rainfall was 1·354 inches, distributed over 15 days. The average rainfall for December in the twenty-three years, 1865–87, inclusive, was 2·419 inches, and the average number of rainy days was 17·0. The rainfall, therefore, was decidedly below the average, while the rainy days were also below it. In 1876 the rainfall in December was very large—7·566 inches on 22 days. In 1872, 4·682 inches fell on as many as 24 days; and in 1868 (which was otherwise a fine and dry year), 4·740 inches fell on as many as 27 days. On the other hand, in 1867, only ·771 of an inch was measured on 13 days; and in 1871, the December rainfall was only ·797 of an inch on 15 days. In 1885, only ·742 of an inch of rain was measured on but 10 days, but in 1886 the rainfall was 3·348 inches, distributed over as many as 21 days. In 1887 ("the dry year"), the rainfall was 1·323 inches on 10 days; in 1888, it was 2·011 inches on 17 days.

A lunar halo appeared on the 31st. High winds were noted on 13 days, and attained the force of a gale on only two occasions, the 12th and the 13th. The atmosphere was more or less foggy in Dublin on the 5th, 6th, 14th, 15th, 21st, 29th, and 30th. Lightning was seen on the night of the 12th. A little sleet fell on the 25th.

Almost throughout the first week (1st–7th, inclusive) atmospherical pressure was abnormally high over Europe. On Thursday, the 5th, indeed, the barometer read 31·00 inches and upwards in Central Russia, and exceeded 30·50 inches all over Scandinavia, Northern Germany, and the British Islands. Considerable cold existed within the limits of this vast anticyclone. It was most intense in a band stretching from the north and centre of Russia to Sweden as well as through Germany to France. At the beginning of the week the weather was dull, mild, and windy in Ireland, which was under the influence of atmospherical depressions passing northwards across the Atlantic; while calms or light S.E. to E. winds and sharp frosts prevailed in England. On Tuesday temperature gave way in Ireland also, and frosts occurred on Thursday and Friday. During the afternoon of the latter day a S.W. wind freshened, temperature rose, and rain set in. Saturday was fair, bright, and cold in Ireland, snowy in England. In Dublin the mean height of the barometer was 30·395 inches—pressure ranging between 30·016 inches at 9 a.m. of Sunday (wind, S.) and 30·730 inches at 9 p.m. of Thursday (wind, also S.). The mean dry bulb temperature at 9 a.m. and 9 p.m. was 40·8°. The mean of the daily highest and lowest shade temperatures was 41·7°. The thermometer rose to 48·0° on Sunday and fell to 30·6° on Friday. Rain fell on Friday to the amount of ·[illegible] inch; and on the morning of Sunday, the 8th, to the amount of ·[illegible] inch—total measurement, ·070 inch, on two days.

Particularly changeable weather prevailed in all districts during the second week (8th–14th inclusive) pressure, temperature, and wind being variable beyond measure. Several extensive and even deep depressions travelled north-eastwards across the Atlantic, British Isles, and Scandinavia—as these systems approached, the wind became southerly, temperature rose, clouds increased, and rain fell; as they passed away the wind veered to W. and N.W. and fell light, with clearing skies and frost. Sunday was dull and rainy. Monday was showery with fine intervals. Tuesday and Wednesday were cold and fair. Thursday was at first frosty—then rainy, with squalls; lightning was seen in the evening. Friday was a fine, cold day. Saturday was frosty and foggy in the forenoon; cloudy and milder in the afternoon. In Dublin the mean height of the barometer was 29·907 inches—pressure ranged from 30·226 inches at 9 a.m. of Sunday (wind, S.W.) to 29·266 inches at 3 p.m. of Tuesday (wind, W.N.W.). The mean dry bulb temperature at 9 a.m. and 9 p.m. was 41·2°. The mean of the daily maxima and minima was 41·8°. On Monday the thermometer in the screen rose to 53·0°, on Thursday they fell to 30·9°. The lowest temperature on the grass was 22·9° on Thursday. The rainfall amounted to ·335 inch, on four days. Of this amount, ·180 inch was registered on Sunday and ·100 inch on Thursday. There was a dense smoke fog on the forenoon of Saturday and lightning occurred on Thursday evening.

A south-westerly air-current and open unsettled weather prevailed throughout the third week ending Saturday, the 21st. Atmospherical pressure was persistently high over France and Germany, low to the north-westward of the British Isles and over Norway. In the area covered by the high pressure, calms or variable winds and hard frost prevailed. In the British Isles and Scandinavia, on the contrary, squally winds between S.W. and N.W., and frequent showers, with but little frost, were prevalent. Sunday was a cool foggy day in Dublin—the max. temp. being only 41·8°. Monday was mild and damp, and at night the thermometer rose to 56°. The next three days were squally; on Thursday evening rain fell heavily all over Ireland and in parts of Great Britain later on. Friday was clear and cold. Saturday—at first fine—proved wet and dull in the afternoon. In Dublin the mean height of the barometer was 30·053 inches—pressure ranging from 30·403 inches at 9 a.m. of Sunday (wind, W., light) to 29·461 inches at 9 p.m. of Saturday (wind, S.E.). The mean dry bulb temperature at 9 a.m. and 9 p.m. was 46·8°. The mean of the daily max. and min. readings of the screen thermometers was 46·7°. The absolute maximum was 56·0° on Tuesday; the absolute minimum was 34·6° on Sunday. Rain fell on five days to the amount of ·626 inch—the heaviest falls in 24 hours being ·274 inch on Thursday and ·231 inch on Saturday. There was a very dense fog over the city on Sunday.

During the week ended Saturday, the 28th, the distribution of atmospherical pressure was of a mixed type—at first chiefly cyclonic, especially in the far N. and N.W.; afterwards anticyclonic. The weather was generally open, and on Christmas Day frost was absent from the British Isles, France, Western Germany, and the greater part of Scandinavia. This was the more remarkable since the 22nd and 23rd very low temperatures were recorded from the Gulf of Bothnia, the 8 a.m. readings on those days at Haparanda being + 7° and —4° F. respectively. Up to and including Thursday, the 26th, rain fell very generally—the weather afterwards becoming unusually dry, mild and quiet. In Dublin, Sunday was dull and damp. Monday showed an improvement, but at night there were squalls and rain. Christmas Eve was bright and bracing, and three beautiful days followed—mild, sunny, and dry, quite [illegible]. Saturday was dull, mild, and showery, with light variable winds. In Dublin the mean height of the barometer was [illegible] inches, pressure ranging from [illegible] inches at 9 a.m. of Sunday (wind, S.W.) to 30.473 inches at 9 p.m. of Thursday (wind, S.S.W.). At 8 a.m. next morning (Friday) readings as high as [illegible] inches at Wisby, in Gothland, and [illegible] inches at Stockholm, were recorded. In Dublin the mean dry bulb temperature at 9 a.m. and 9 p.m. was [illegible]°. The mean of the daily highest and lowest temperatures was [illegible]°. The screened thermometers rose to [illegible]° in the night of the 23rd–24th, and fell to 37.4° on Saturday. The rainfall amounted to [illegible] inch, and was distributed over three days—[illegible] inch fell on Monday, and [illegible] inch on Saturday.

The last three days of the month were mild and chiefly fine in Dublin, while [illegible] frost prevailed in the S.E. and centre of England.

The rainfall in Dublin during the year ending December 31st amounted to 27.178 inches on 193 days, compared with [illegible] inches on 190 days in 1888, 16.001 inches on 160 days in 1887, and a twenty-three years' average of 27.678 inches on 194.4 days.

At Greystones, Co. Wicklow, the rainfall in December, 1889, was [illegible] inches distributed over 13 days. Of this quantity [illegible] of an inch fell on the 25th, [illegible] of an inch on the 18th, and [illegible] of an inch on the 19th. During the year 31.260 inches of rain fell at Greystones, on, however, only 132 days. In 1888 the corresponding figures were 34.240 inches of rain on [illegible] days.

Rainfall in 1889,

At 40, Fitzwilliam-square, West, Dublin.

Rain Gauge:—Diameter of funnel, 8 in. Height of top—Above ground, [illegible] ft. [illegible] in.; above sea level, [illegible] ft.

Month.	Total Depth.	Greatest Fall in 24 hours.		Number of Days on which .01 or more fell.	Month.	Total Depth.	Greatest Fall in 24 hours.		Number of Days on which .01 or more fell.
	Inches.	Depth.	Date.			Inches.	Depth.	Date.	
January,	[illegible]	[illegible]	[illegible]	[illegible]	August,	[illegible]	[illegible]	[illegible]	[illegible]
February,	[illegible]	[illegible]	[illegible]	[illegible]	September,	[illegible]	[illegible]	[illegible]	[illegible]
March,	[illegible]	[illegible]	[illegible]	[illegible]	October,	[illegible]	[illegible]	[illegible]	[illegible]
April,	[illegible]	[illegible]	[illegible]	[illegible]	November,	[illegible]	[illegible]	[illegible]	[illegible]
May,	[illegible]	[illegible]	[illegible]	[illegible]	December,	[illegible]	[illegible]	[illegible]	[illegible]
June,	[illegible]	[illegible]	[illegible]	[illegible]					
July,	[illegible]	[illegible]	[illegible]	[illegible]	Total,	[illegible]	-	-	[illegible]

The rainfall was five-tenths of an inch in defect of the average annual measurement of the twenty-three years, 1865–87, inclusive—viz., 27.678 inches.

It will be remembered that the rainfall in 1887 was very exceptionally small—16.001 inches, the only approach to this measurement in Dublin being in 1870, when only [illegible] inches fell, and in 1884, when the measurement was [illegible] inches. In seven of the twenty-three years in question the rainfall was less than 26 inches, and in 1883 it was 26.614 inches.

The scanty rainfall in 1887 was in marked contrast to the abundant downpour in 1886, when [illegible] inches—or as nearly as possible double the fall of 1887—fell on 220 days. Only twice since these records commenced has the rainfall in Dublin exceeded that of 1886—namely, in 1872, when [illegible] inches fell on [illegible] days, and in 1880, when [illegible] inches were measured on, however, only [illegible] days.

In 1889, there were 193 rainy days, or days upon which not less than .01 inch of rain (one hundredth of an inch) was measured. This was slightly in defect of the average number of rainy days, which was 194.4 in the twenty-three years, 1865–87, inclusive. In 1868—the warm dry year of recent [illegible]—as well as in 1887, the rainy days were only 1[illegible], and in 1870 they were only 14[illegible]. In 18[illegible], however, the rainfall amounted to 2[illegible] inches, or more than [illegible] inches above the [illegible]. In 1887, and even in 1870, [illegible] inches were recorded. Included in the 1[illegible] rainy days in 18[illegible] are 1[illegible] on which snow or sleet fell, and [illegible] on which there was hail. In February hail was observed on 7 days, in March on [illegible] days, and in April on [illegible] days. Hail also fell twice in January and October, and once in May, July, and November. Snow or sleet fell on 1 day in January, on [illegible] days in February, on [illegible] days in March, on 1 day in April, on 2 days in November, and on 1 day in December. Thunder and lightning occurred on five occasions during the year—once in April, July, and October, and twice in August. Lightning was also seen on one occasion in each of the last three months of the year.

The rainfall was distributed as follows:—[illegible] inches fell on 53 days in the first quarter, [illegible] inches on [illegible] days in the second, 9·360 inches on 50 days in the third, and 7·[illegible] inches on [illegible] days in the fourth and last quarter.

Of the 9·360 inches which fell in the third quarter of the year, 5·747 inches were measured in August.

Abstract of Meteorological Observations taken at Dublin (40, Fitzwilliam-square, West), during the Year 18[illegible].

Month.	Abs. Max.	Date.	Abs. Min.	Date.	Mean Daily Max.	Mean Daily Min.	Rainfall.	Rainy Days.	Mean Height of Barometer.	Highest Pressure.	Date.	Lowest Pressure.	Date.	Prevalent Winds.
	°		°		°	°	Ins.		Ins.	Ins.		Ins.		
[illegible]	[illegible]	[illegible]	[illegible]	[illegible]	[illegible]	[illegible]	[illegible]	[illegible]	[illegible]	[illegible]	[illegible]	[illegible]	[illegible]	[illegible]
[illegible]	[illegible]	[illegible]	[illegible]	[illegible]	[illegible]	[illegible]	[illegible]	[illegible]	[illegible]	[illegible]	[illegible]	[illegible]	[illegible]	[illegible]
[illegible]	[illegible]	[illegible]	[illegible]	[illegible]	[illegible]	[illegible]	[illegible]	[illegible]	[illegible]	[illegible]	[illegible]	[illegible]	[illegible]	[illegible]
[illegible]	[illegible]	[illegible]	[illegible]	[illegible]	[illegible]	[illegible]	[illegible]	[illegible]	[illegible]	[illegible]	[illegible]	[illegible]	[illegible]	[illegible]
[illegible]	[illegible]	[illegible]	[illegible]	[illegible]	[illegible]	[illegible]	[illegible]	[illegible]	[illegible]	[illegible]	[illegible]	[illegible]	[illegible]	[illegible]
[illegible]	[illegible]	[illegible]	[illegible]	[illegible]	[illegible]	[illegible]	[illegible]	[illegible]	[illegible]	[illegible]	[illegible]	[illegible]	[illegible]	[illegible]
[illegible]	[illegible]	[illegible]	[illegible]	[illegible]	[illegible]	[illegible]	[illegible]	[illegible]	[illegible]	[illegible]	[illegible]	[illegible]	[illegible]	[illegible]
[illegible]	[illegible]	[illegible]	[illegible]	[illegible]	[illegible]	[illegible]	[illegible]	[illegible]	[illegible]	[illegible]	[illegible]	[illegible]	[illegible]	[illegible]
[illegible]	[illegible]	[illegible]	[illegible]	[illegible]	[illegible]	[illegible]	[illegible]	[illegible]	[illegible]	[illegible]	[illegible]	[illegible]	[illegible]	[illegible]
[illegible]	[illegible]	[illegible]	[illegible]	[illegible]	[illegible]	[illegible]	[illegible]	[illegible]	[illegible]	[illegible]	[illegible]	[illegible]	[illegible]	[illegible]
[illegible]	[illegible]	[illegible]	[illegible]	[illegible]	[illegible]	[illegible]	[illegible]	[illegible]	[illegible]	[illegible]	[illegible]	[illegible]	[illegible]	[illegible]

TABLE showing the Monthly and Yearly Rainfall at Dublin during the Twenty-one Years 1869 to 1889, inclusive; with the Means for the Twenty Years 1869 to 1888.

Year.	January.	February.	March.	April.	May.	June.	July.	August.	September.	October.	November.	December.	Total Rainfall.
	Inches.	Inches.	Inches.	Inches.	Inches.	Inches.	Inches.	Inches.	Inches.	Inches.	Inches.	Inches.	Inches.
1869,	[illegible]	[illegible]	[illegible]	[illegible]	[illegible]	[illegible]	[illegible]	[illegible]	[illegible]	[illegible]	[illegible]	[illegible]	[illegible]
1870,	[illegible]	[illegible]	[illegible]	[illegible]	[illegible]	[illegible]	[illegible]	[illegible]	[illegible]	[illegible]	[illegible]	[illegible]	[illegible]
1871,	[illegible]	[illegible]	[illegible]	[illegible]	[illegible]	[illegible]	[illegible]	[illegible]	[illegible]	[illegible]	[illegible]	[illegible]	[illegible]
1872,	[illegible]	[illegible]	[illegible]	[illegible]	[illegible]	[illegible]	[illegible]	[illegible]	[illegible]	[illegible]	[illegible]	[illegible]	[illegible]
1873,	[illegible]	[illegible]	[illegible]	[illegible]	[illegible]	[illegible]	[illegible]	[illegible]	[illegible]	[illegible]	[illegible]	[illegible]	[illegible]
1874,	[illegible]	[illegible]	[illegible]	[illegible]	[illegible]	[illegible]	[illegible]	[illegible]	[illegible]	[illegible]	[illegible]	[illegible]	[illegible]
1875,	[illegible]	[illegible]	[illegible]	[illegible]	[illegible]	[illegible]	[illegible]	[illegible]	[illegible]	[illegible]	[illegible]	[illegible]	[illegible]
1876,	[illegible]	[illegible]	[illegible]	[illegible]	[illegible]	[illegible]	[illegible]	[illegible]	[illegible]	[illegible]	[illegible]	[illegible]	[illegible]
1877,	[illegible]	[illegible]	[illegible]	[illegible]	[illegible]	[illegible]	[illegible]	[illegible]	[illegible]	[illegible]	[illegible]	[illegible]	[illegible]
1878,	[illegible]	[illegible]	[illegible]	[illegible]	[illegible]	[illegible]	[illegible]	[illegible]	[illegible]	[illegible]	[illegible]	[illegible]	[illegible]
1879,	[illegible]	[illegible]	[illegible]	[illegible]	[illegible]	[illegible]	[illegible]	[illegible]	[illegible]	[illegible]	[illegible]	[illegible]	[illegible]
1880,	[illegible]	[illegible]	[illegible]	[illegible]	[illegible]	[illegible]	[illegible]	[illegible]	[illegible]	[illegible]	[illegible]	[illegible]	[illegible]
1881,	[illegible]	[illegible]	[illegible]	[illegible]	[illegible]	[illegible]	[illegible]	[illegible]	[illegible]	[illegible]	[illegible]	[illegible]	[illegible]
1882,	[illegible]	[illegible]	[illegible]	[illegible]	[illegible]	[illegible]	[illegible]	[illegible]	[illegible]	[illegible]	[illegible]	[illegible]	[illegible]
1883,	[illegible]	[illegible]	[illegible]	[illegible]	[illegible]	[illegible]	[illegible]	[illegible]	[illegible]	[illegible]	[illegible]	[illegible]	[illegible]
1884,	[illegible]	[illegible]	[illegible]	[illegible]	[illegible]	[illegible]	[illegible]	[illegible]	[illegible]	[illegible]	[illegible]	[illegible]	[illegible]
1885,	[illegible]	[illegible]	[illegible]	[illegible]	[illegible]	[illegible]	[illegible]	[illegible]	[illegible]	[illegible]	[illegible]	[illegible]	[illegible]
1886,	[illegible]	[illegible]	[illegible]	[illegible]	[illegible]	[illegible]	[illegible]	[illegible]	[illegible]	[illegible]	[illegible]	[illegible]	[illegible]
1887,	[illegible]	[illegible]	[illegible]	[illegible]	[illegible]	[illegible]	[illegible]	[illegible]	[illegible]	[illegible]	[illegible]	[illegible]	[illegible]
1888,	[illegible]	[illegible]	[illegible]	[illegible]	[illegible]	[illegible]	[illegible]	[illegible]	[illegible]	[illegible]	[illegible]	[illegible]	[illegible]
Means,	[illegible]	[illegible]	[illegible]	[illegible]	[illegible]	[illegible]	[illegible]	[illegible]	[illegible]	[illegible]	[illegible]	[illegible]	[illegible]
1889,	[illegible]	[illegible]	[illegible]	[illegible]	[illegible]	[illegible]	[illegible]	[illegible]	[illegible]	[illegible]	[illegible]	[illegible]	[illegible]

[illegible footnotes]

TABLE showing the Monthly and Yearly Number of Rainy Days* at Dublin during the Twenty-one Years 1869 to 1889, inclusive; with the Means for the Twenty Years 1869 to 1888.

Year.	January.	February.	March.	April.	May.	June.	July.	August.	September.	October.	November.	December.	Total Rainy Days.
1869,	[illegible]	[illegible]	[illegible]	[illegible]	[illegible]	[illegible]	[illegible]	[illegible]	[illegible]	[illegible]	[illegible]	[illegible]	[illegible]
1870,	[illegible]	[illegible]	[illegible]	[illegible]	[illegible]	[illegible]	[illegible]	[illegible]	[illegible]	[illegible]	[illegible]	[illegible]	[illegible]
1871,	[illegible]	[illegible]	[illegible]	[illegible]	[illegible]	[illegible]	[illegible]	[illegible]	[illegible]	[illegible]	[illegible]	[illegible]	[illegible]
1872,	[illegible]	[illegible]	[illegible]	[illegible]	[illegible]	[illegible]	[illegible]	[illegible]	[illegible]	[illegible]	[illegible]	[illegible]	[illegible]
1873,	[illegible]	[illegible]	[illegible]	[illegible]	[illegible]	[illegible]	[illegible]	[illegible]	[illegible]	[illegible]	[illegible]	[illegible]	[illegible]
1874,	[illegible]	[illegible]	[illegible]	[illegible]	[illegible]	[illegible]	[illegible]	[illegible]	[illegible]	[illegible]	[illegible]	[illegible]	[illegible]
1875,	[illegible]	[illegible]	[illegible]	[illegible]	[illegible]	[illegible]	[illegible]	[illegible]	[illegible]	[illegible]	[illegible]	[illegible]	[illegible]
1876,	[illegible]	[illegible]	[illegible]	[illegible]	[illegible]	[illegible]	[illegible]	[illegible]	[illegible]	[illegible]	[illegible]	[illegible]	[illegible]
1877,	[illegible]	[illegible]	[illegible]	[illegible]	[illegible]	[illegible]	[illegible]	[illegible]	[illegible]	[illegible]	[illegible]	[illegible]	[illegible]
1878,	[illegible]	[illegible]	[illegible]	[illegible]	[illegible]	[illegible]	[illegible]	[illegible]	[illegible]	[illegible]	[illegible]	[illegible]	[illegible]
1879,	[illegible]	[illegible]	[illegible]	[illegible]	[illegible]	[illegible]	[illegible]	[illegible]	[illegible]	[illegible]	[illegible]	[illegible]	[illegible]
1880,	[illegible]	[illegible]	[illegible]	[illegible]	[illegible]	[illegible]	[illegible]	[illegible]	[illegible]	[illegible]	[illegible]	[illegible]	[illegible]
1881,	[illegible]	[illegible]	[illegible]	[illegible]	[illegible]	[illegible]	[illegible]	[illegible]	[illegible]	[illegible]	[illegible]	[illegible]	[illegible]
1882,	[illegible]	[illegible]	[illegible]	[illegible]	[illegible]	[illegible]	[illegible]	[illegible]	[illegible]	[illegible]	[illegible]	[illegible]	[illegible]
1883,	[illegible]	[illegible]	[illegible]	[illegible]	[illegible]	[illegible]	[illegible]	[illegible]	[illegible]	[illegible]	[illegible]	[illegible]	[illegible]
1884,	[illegible]	[illegible]	[illegible]	[illegible]	[illegible]	[illegible]	[illegible]	[illegible]	[illegible]	[illegible]	[illegible]	[illegible]	[illegible]
1885,	[illegible]	[illegible]	[illegible]	[illegible]	[illegible]	[illegible]	[illegible]	[illegible]	[illegible]	[illegible]	[illegible]	[illegible]	[illegible]
1886,	[illegible]	[illegible]	[illegible]	[illegible]	[illegible]	[illegible]	[illegible]	[illegible]	[illegible]	[illegible]	[illegible]	[illegible]	[illegible]
1887,	[illegible]	[illegible]	[illegible]	[illegible]	[illegible]	[illegible]	[illegible]	[illegible]	[illegible]	[illegible]	[illegible]	[illegible]	[illegible]
1888,	[illegible]	[illegible]	[illegible]	[illegible]	[illegible]	[illegible]	[illegible]	[illegible]	[illegible]	[illegible]	[illegible]	[illegible]	[illegible]
Means,	[illegible]	[illegible]	[illegible]	[illegible]	[illegible]	[illegible]	[illegible]	[illegible]	[illegible]	[illegible]	[illegible]	[illegible]	[illegible]
1889,	[illegible]	[illegible]	[illegible]	[illegible]	[illegible]	[illegible]	[illegible]	[illegible]	[illegible]	[illegible]	[illegible]	[illegible]	[illegible]

[illegible footnotes]

TABLE showing the Temperature of the Air in Dublin in 1889, and the Average Temperature for the Twenty Years 1869 to 1888, inclusive, as recorded by Dr. J. W. Moore.

Year.	January.	February.	March.	April.	May.	June.	July.	August.	September.	October.	November.	December.	Year.
1869, ...	[illegible]	[illegible]	[illegible]	[illegible]	[illegible]	[illegible]	[illegible]	[illegible]	[illegible]	[illegible]	[illegible]	[illegible]	[illegible]
1870, ...	[illegible]	[illegible]	[illegible]	[illegible]	[illegible]	[illegible]	[illegible]	[illegible]	[illegible]	[illegible]	[illegible]	[illegible]	[illegible]
1871, ...	[illegible]	[illegible]	[illegible]	[illegible]	[illegible]	[illegible]	[illegible]	[illegible]	[illegible]	[illegible]	[illegible]	[illegible]	[illegible]
1872, ...	[illegible]	[illegible]	[illegible]	[illegible]	[illegible]	[illegible]	[illegible]	[illegible]	[illegible]	[illegible]	[illegible]	[illegible]	[illegible]
1873, ...	[illegible]	[illegible]	[illegible]	[illegible]	[illegible]	[illegible]	[illegible]	[illegible]	[illegible]	[illegible]	[illegible]	[illegible]	[illegible]
1874, ...	[illegible]	[illegible]	[illegible]	[illegible]	[illegible]	[illegible]	[illegible]	[illegible]	[illegible]	[illegible]	[illegible]	[illegible]	[illegible]
1875, ...	[illegible]	[illegible]	43·1	[illegible]	[illegible]	[illegible]	[illegible]	[illegible]	[illegible]	[illegible]	[illegible]	[illegible]	[illegible]
1876, ...	[illegible]	[illegible]	[illegible]	[illegible]	[illegible]	[illegible]	[illegible]	[illegible]	[illegible]	[illegible]	[illegible]	[illegible]	[illegible]
1877, ...	[illegible]	[illegible]	[illegible]	[illegible]	[illegible]	[illegible]	[illegible]	[illegible]	[illegible]	[illegible]	[illegible]	[illegible]	[illegible]
1878, ...	[illegible]	[illegible]	[illegible]	[illegible]	[illegible]	[illegible]	[illegible]	[illegible]	[illegible]	[illegible]	[illegible]	[illegible]	[illegible]
1879, ...	[illegible]	[illegible]	[illegible]	[illegible]	[illegible]	[illegible]	[illegible]	[illegible]	[illegible]	[illegible]	[illegible]	[illegible]	[illegible]
1880, ...	[illegible]	[illegible]	[illegible]	[illegible]	[illegible]	[illegible]	[illegible]	[illegible]	[illegible]	[illegible]	[illegible]	[illegible]	[illegible]
1881, ...	[illegible]	[illegible]	[illegible]	[illegible]	[illegible]	[illegible]	[illegible]	[illegible]	[illegible]	[illegible]	[illegible]	[illegible]	[illegible]
1882, ...	[illegible]	[illegible]	[illegible]	[illegible]	[illegible]	[illegible]	[illegible]	[illegible]	[illegible]	[illegible]	[illegible]	[illegible]	[illegible]
1883, ...	[illegible]	[illegible]	[illegible]	[illegible]	[illegible]	[illegible]	[illegible]	[illegible]	[illegible]	[illegible]	[illegible]	[illegible]	[illegible]
1884, ...	[illegible]	[illegible]	[illegible]	[illegible]	[illegible]	[illegible]	[illegible]	[illegible]	[illegible]	[illegible]	[illegible]	[illegible]	[illegible]
1885, ...	[illegible]	[illegible]	[illegible]	[illegible]	[illegible]	[illegible]	[illegible]	[illegible]	[illegible]	[illegible]	[illegible]	[illegible]	[illegible]
1886, ...	[illegible]	[illegible]	[illegible]	[illegible]	[illegible]	[illegible]	[illegible]	[illegible]	[illegible]	[illegible]	[illegible]	[illegible]	[illegible]
1887, ...	[illegible]	[illegible]	[illegible]	[illegible]	[illegible]	[illegible]	[illegible]	[illegible]	[illegible]	[illegible]	[illegible]	[illegible]	[illegible]
1888, ...	[illegible]	[illegible]	[illegible]	[illegible]	[illegible]	[illegible]	[illegible]	[illegible]	[illegible]	[illegible]	[illegible]	[illegible]	[illegible]
Average, .	[illegible]	[illegible]	[illegible]	[illegible]	[illegible]	[illegible]	[illegible]	[illegible]	[illegible]	[illegible]	[illegible]	[illegible]	[illegible]
1889, ...	[illegible]	[illegible]	[illegible]	[illegible]	[illegible]	[illegible]	[illegible]	[illegible]	[illegible]	[illegible]	[illegible]	[illegible]	[illegible]

N.B.—The temperatures given above were deduced from the maximal and minimal readings of the Thermometer by [illegible] Formula, viz.— [illegible] = Mean Temperature.

Dublin Castle,
8th July, 1890.

Sir,

I have to acknowledge the receipt of your letter of the 7th instant, forwarding, for submission to His Excellency the Lord Lieutenant, the Agricultural Statistics of Ireland for the year 1889.

I am, Sir,
Your obedient Servant,
W. Ridgeway.

The Registrar-General,
Charlemont House,
Rutland-square.

Dublin: Printed for Her Majesty's Stationery Office,
By Alexr. Thom & Co. (Limited), 87, 88, and 89, Abbey-street.
The Queen's Printing Office.

www.ingramcontent.com/pod-product-compliance
Lightning Source LLC
Chambersburg PA
CBHW021814190326
41518CB00007B/597

* 9 7 8 3 7 4 2 8 0 0 2 4 4 *